T0388689

Palgrave Studies in the History of Economic Thought

Palgrave Studies in the History of Economic Thought publishes contributions by leading scholars, illuminating key events, theories and individuals that have had a lasting impact on the development of modern-day economics. The topics covered include the development of economies, institutions and theories.

More information about this series at
http://www.palgrave.com/gp/series/14585

Fiorenzo Mornati

Vilfredo Pareto: An Intellectual Biography Volume III

From Liberty to Science (1898–1923)

palgrave
macmillan

Fiorenzo Mornati
Dipto di Econ e Statistica
University of Turin
Torino, Italy

ISSN 2662-6578 ISSN 2662-6586 (electronic)
Palgrave Studies in the History of Economic Thought
ISBN 978-3-030-57756-8 ISBN 978-3-030-57757-5 (eBook)
https://doi.org/10.1007/978-3-030-57757-5

Cover illustration: INTERFOTO / Alamy Stock Photo

This Palgrave Macmillan imprint is published by the registered company Springer Nature Switzerland AG.
The registered company address is: Gewerbestrasse 11, 6330 Cham, Switzerland

Preface

In this volume, the third of the trilogy, we shall deal with the final, very intense period in Pareto's intellectual biography when, by now largely free of political distractions, he was able to further investigate and to attempt a synthesis of the two disciplines to which he had devoted most attention, economics and sociology.

Considering pure economics to be now capable of autonomous development, that is free of the involvement of specific external factors, we shall initially turn our attention to the definitive conclusions reached by Pareto with particular regard to the architecture of general equilibrium, based exclusively on the empirical factor represented by the curve of indifference.

The final form of Pareto's sociology appears, instead, to spring from the disenchantment he now felt in observing the irreversible decline of liberalism together with the apparently irresistible rise of socialism. These two phenomena, with their apparent negation of the logic whereby societies will constantly seek to attain maximum well-being, appear to have stimulated Pareto to attempt the formulation of the comprehensive account appearing in the *Treatise* on general sociology, of which we will examine the salient features. The war and the post-war period provided Pareto with an opportunity to perform a—somewhat self-congratulatory—verification of the plausibility of the sociological model he had developed.

Again, in this volume we will draw on the resources found in the Paretology in a highly selective manner, presenting the formal arguments in as complete but accessible a manner as possible.

We extend our thanks to Roberto Marchionatti for our exchanges on a variety of topics relating also to this volume and to the following for their valuable and patient collaboration in archive research (in alphabetical order): the Banca Popolare di Sondrio (owners of the Vilfredo Pareto letter archive held at the Luigi Credaro library in Sondrio), Piercarlo Della Ferrera (custodian of the archive), the Interlibrary loans service of the Norberto Bobbio Library at the University of Turin and the Vaud Cantonal Archives in Lausanne.

Torino, Italy Fiorenzo Mornati

Contents

CHAPTER 1

A New Pure Economics

In the decade following his address at the "Stella" student association in Lausanne and culminating in the publication of the French version of his *Manual of Political Economy*, Pareto's conception of theoretical (or "pure") economics, hitherto constituting an introduction to the field of applied economics which followed broadly in the footsteps of Walras,[1] appears to have undergone a definitive and largely self-sustained development which was distinctly original, even if never disconnected from Pareto's other interests in the social sciences.[2] Hence, in this chapter, we will describe his ground-breaking theory of choice (Sects. 1.1 and 1.2) as well as the aspects of the *Manual* which display innovations in relation to Pareto's economic thinking of the immediately preceding period (Sect. 1.3). We will then characterise the definitive description of pure economics offered by Pareto (Sect. 1.4) together with a selection of his critical—and self-critical—remarks on the discipline (Sect. 1.5).

1.1 The Beginnings of Paretian Pure Economics: Early References to the Theory of Choice

On the 14th of December 1898[3] Pareto informed Pantaleoni that for the meeting of the Lausanne student association "Stella" on the following 17th of December, when he was due to be nominated an honorary member,[4] he had prepared the paper *Comment se pose le problème de l'économie*

F. Mornati, *Vilfredo Pareto: An Intellectual Biography Volume III*,
Palgrave Studies in the History of Economic Thought,
https://doi.org/10.1007/978-3-030-57757-5_1

pure (Expounding the pure economics). In this paper he showed "how to circumvent the difficulty arising from the impossibility of measuring ophelimity". This is recognised as constituting a decisive step forward in the development of Pareto's theory of utility which was formalised with the abandoning of the always problematical cardinal conception of utility in favour of an ordinal alternative, at least for theoretical purposes. In his paper Pareto affirmed that in order to construct a pure economics, it was necessary, firstly, to conceive the economic characteristic of mankind to be "the pursuit of pleasure and the avoidance of pain"; secondly, to conceive of both pleasure and pain as quantities and, thirdly, to bear in mind that recognising the existence of quantities and actually measuring them are two different things.[5]

Consequently, Pareto defined pure economics as "a type of rational mechanics" dealing not with points but with *homines oeconomici* ("*economic agents*").[6] If one of these agents possesses variable quantities q_a and q_b of commodities A e B, each possible combination of these quantities "will generate different degrees of utility".[7]

Thus, if we are able to measure the separate increases in utility

$$\varphi_a d_a \text{ and } \varphi_a d_a$$

which the economic agent will obtain in passing from q_a to dq_a (or similarly from q_b to dq_b), then the agent will enjoy an overall increase in utility equal to

$$\varphi_a dq_a + \varphi_b dq_b$$

which will be maximised with the occurrence of the quantities of A and B such that[8]

$$\varphi_a dq_a + \varphi_b dq_b = 0$$

If we further posit that "the utility is independent of the order of consumption",[9] then φ_a and φ_b constitute the first-order partial derivatives of a function Φ[10] and so the previous equation can be substituted by[11]

$$d\Phi = 0$$

In reality, φ_a and φ_b cannot be measured, and therefore neither can Φ. But every human being knows, "with certainty, since this is a matter of logical actions", whether, in passing from one combination to the next, his utility increases or diminishes, that is, whether $d\Phi$ increases or diminishes.[12]

A further Ψ function can also be constructed "whose values, while partially arbitrary, are such that $d\Psi$ will always have the same sign as $d\Phi$".[13]

Hence,

$$d\Phi = 0$$

can be replaced with

$$d\psi = 0$$

Pareto emphasises[14] that the economic agent's passage from one combination to another of quantities of goods is not unrestricted but is constrained by the conditions of production which, if the goods can be transformed in a fixed proportion, can be represented graphically by a negatively inclined curve and in algebraic terms by the function

$$F(q_a, q_b) = 0$$

which can be specified as

$$a - q_a + (b - q_b) = 0$$

where a and b are the quantities initially possessed of A and B; q_a and q_b are the quantities possessed after the transformation; $a - q_a$ and $b - q_b$ are the quantities in which the transformation of A to B is manifested and α is the constant which represents the fixed ratio at which a quantity of A can be transformed into a quantity of B.

Shortly afterwards Pareto took the opportunity of a brief but seminal exchange of correspondence with the French mathematician Hermann Laurent[15] to clarify his new theory. He agreed with his correspondent that only "elements which are capable of equivalence and addition" can be measured, but added that, in his view, utility possesses these qualities on an empirical level, since "each day, we weigh one type of pleasure against another, we judge them to be equal, greater or lesser". Nevertheless,

Pareto was also prepared to admit that the possession of such qualities represented simply a postulate, a procedure to which, for that matter, every science resorts in its infancy.[16]

In the light of this, Pareto specified that this general problem of economics implies the existence of the equation[17]

$$d\varphi = \varphi_a dx_a + \varphi_b dx_{ab}$$

where the value of φ "depends on the order of consumption"[18] while x_a, x_b, ... represent "the successive values of the quantities exchanged".[19]

Further, Pareto observed that in the differential of what we now call the budget constraint

$$p_a dx_a + p_b dx_{ab} + \ldots = 0$$

the prices do not depend on x_a, x_b ... but on the values which the quantities of goods have at the conclusion of the exchange (designated r_a, r_b ... by Pareto) so that the integral of the equation at the end of the exchange is

$$p_a r_a + p_b r_{ab} + \ldots = 0$$

representing "simply the statement of condition of the individual concerned: income = expenditure", which does not thus permit the determination of the quantities of goods exchanged.[20]

These quantities are in fact the product of the system consisting of

$$d\varphi = \varphi_a dx_a + \varphi_b dx_b + \ldots$$

and of the n - 1 equations

$$p_a \left(\frac{dx_a}{dx_b} \right) + p_b = 0; p_a \left(\frac{dx_a}{dx_c} \right) + p_c = 0; \ldots.$$

which are obtained from

$$p_a dx_a + p_b dx_b + \ldots = 0$$

bearing in mind that the latter establishes "a relation between the n quantities" x_a, x_b ..., with the implication that one of these, for example x_a itself, can be considered a function of the remaining n - 1.[21]

In geometric terms, if the quantities r_a and r_a are connected by the function

$$\varphi(r_a, r_b) = \text{constant}$$

obtainable only "through experience",[22] this latter can be represented on the plane "of the contour lines of any other surface"[23]

$$\varphi_a dr_a + \varphi_b dr_b = 0.$$

The equation

$$p_a dx_a + p_b dx_b = 0$$

(which is a straight line, if the prices do not depend on the quantities x_a, x_b) itself represents "the differential equation of a path followed during the exchange ... [which] stops when the path is tangential to a contour line",[24] where it assumes the form

$$p_a dr_a + p_b dr_b = 0$$

Pareto then reiterates that in order to achieve the condition of tangency of the straight line to the contour lines, both the equation of the straight line and the equation of a contour line are necessary and hence the system[25]

$$\varphi_a dr_a + \varphi_b dr_b = 0$$
$$p_a dr_a + p_b dr_b = 0$$

whose solution is yielded by the combination of r_a and r_b such that

$$\frac{\varphi_a}{\varphi_b} = -\left(\frac{dr_a}{dr_b}\right) = \frac{p_a}{p_b}$$

or such that

$$\frac{\varphi_a}{p_a} = \frac{\varphi_b}{p_b}$$

Pareto deduces that, as also "generally in practice",[26]

$$p_a = f_a\left(r_a, r_b, \ldots\right); p_b = f_{ab}\left(r_a, r_b, \ldots\right); \ldots$$

Pareto explained to Pantaleoni that while "Edgeworth and the others start from the notion of final degree of utility to arrive at the determination of the indifference curve", he, and "this is the only novelty", "leave[s] completely aside the final degree of utility and start[s] from the curves of indifference which emerge directly from experience".[27] Thus, in the end, in response to "the objections made because the final degree of utility [can] not be measured", we can retort that "there is no need to measure [it]".[28]

Pareto then underlined that for the construction of a curve of indifference, an operation consisting simply in having someone state the different combinations of quantities of two goods which for him are "exactly the same" without asking him "why", there is no "talk of transformations or reasons for transformations or prices" and neither "is it necessary that the elements should be measurable".[29] Moreover, reflecting on a curve of indifference which today we would call standard, that is, continuous, differentiable, strictly convex and negatively inclined, Pareto remarked that "the reason for the exchange or the transformation" between the two commodities "varies constantly":[30] and indeed it is also possible to have curves inclined negatively at right angles (today defined as curves describing lexicographical orders of preference) where "the individual, provided he has *b* of *B*, is indifferent to having any quantity of *A* greater than a".[31] However, positively inclined indifference curves cannot exist because a combination containing greater quantities of both commodities cannot be indifferent in relation to one containing lesser quantities of both.[32]

1.2 The Definitive Formulation of the Theory of Choice

In March 1900, Pareto, proceeding with this reflection, pointed out that "the most basic observation suffices to demonstrate that animals and humans make choices [that] very often [are] constantly repeated", which allows these choices to constitute "the object of a science".[33] Pareto underlined that he was interested exclusively in the fact of the choice, without "in any way" seeking "the reason" for it,[34] which "is only a matter of taste".[35] Pareto also specified that choices "fall on items whose quantities are variable and subject to measurement",[36] with a variability which in practice is discontinuous and which is replaced by a variability which is considered as continuous[37] simply in order to "facilitate the application of the mathematical method". Pareto also thought it advisable to distinguish choices an individual makes on the basis of his own personal preferences[38] from those made "considering the effects they will have on other individuals".[39] Finally, Pareto noted that "in his choices, the individual encounters obstacles"[40] and that he aims to make a choice, referred to as an equilibrium choice, "which he favours over all others, taking into account all the obstacles".[41]

In the light of all this, Pareto claimed that, conceptually, in order to deal with the problem of choice, "the first operation is the creation of a table of the possible choices[42] faced by the individuals under consideration ... indicating their order of preference".[43] In order to proceed, it is necessary to observe the following limitations: "to consider only the choice of economic goods whose quantities are variable and can be measured";[44] to consider "only the state of equilibrium";[45] in the context of a given combination of quantities of goods, to consider "the choice to be indifferent with regard to the order of consumption".[46] Thus, were it not for the obstacles, "the solution to the problem of equilibrium would be very simple. The individual would stop at the point where he was sated with everything".[47] However, the obstacles exist, in the sense that "to obtain certain things he is obliged to forgo others" and therefore it is necessary also to construct "a table of obstacles".[48]

Consequently, the economic problem, described verbally, consists in representing "experience" in the form of a volume on each of whose pages are recorded the combinations which are indifferent among themselves, taking care to "arrange the combinations in order of preference according to the number of pages" and remembering that the obstacles will identify all the pages among which a choice can in fact be made.[49]

In geometric terms, after placing the quantities available of commodities *X* and *Y* on a Cartesian axis, the set of combinations of the quantities *X* and *Y* to which the individual's choice is limited by the obstacles can be represented on a continuous negatively inclined curve.[50] Once this wave curve is drawn,[51] Pareto underlines that the individual, arriving at the curve along "his favoured path", is not obliged to follow it but neither can he pass beyond it.[52] Although for Pareto it is an "exception",[53] the individual may halt at that point neighbouring which there are only points which are dispreferred by him; the "general case", however, is that where the individual stops at a point neighbouring which there are other points which he considers indifferent to the stopping point.[54]

Pareto then introduced the concept of the line of indifference as being the line which "represents types of consumption among which the *economic agent* does not distinguish",[55] while specifying that "only one line of indifference can pass through any given point"[56] and that the set of infinite lines of indifference which can be constructed,[57] each starting from any given combination of quantities of goods is sufficient to "characterise the *economic agent* in regard to all problems of equilibrium".[58]

On this basis, Pareto declared that "the equilibrium position occurs when the curve indicating the effects of the obstacles [and] one of the curves of indifference have a common tangent",[59] which in general is made up of a number of points. The resulting situation of equilibrium "corresponds to the best choice possible, and also to the worst"; which of these is the case is indicated by "the nature of the problem itself".[60]

The graphical representation can be completed on the basis of the following considerations:

to any given combination we can assign "an arbitrary numerical index" and this same index will also be attributed to all the other combinations lying along the line of indifference of which this combination forms part;

let any given combination (such as *A*) on a curve of indifference be compared with another combination *B* which does not form part of the same curve: if *B* is preferred (dispreferred) with regard to *A*, an arbitrary numerical index is assigned to *B* (and to all the combinations lying along the curve of indifference to which it belongs) which, however, is greater (or smaller) than that assigned to *A*;[61]

in this manner we will cover "the plane with an infinite number of lines of indifference which are infinitely close to each other … each [of which] will have an index showing the individual's order of preference",[62] thus generating a "complete representation of the individual's preferences".[63]

Lastly, from the analytical point of view, Pareto stated that "by interpolation", the whole set of lines of indifference can yield the equation

$$G(x,y,\beta)=0 \text{ or } \beta = I(x,y)$$

where the equation of a line of indifference corresponds to each value of the parameter β.[64]

By differentiating

$$\beta = I(x,y)$$

we obtain[65]

$$\left(\frac{\partial I}{\partial x}\right)dx + \left(\frac{\partial I}{\partial y}\right)dy = 0$$

By hypothesising the equation of the obstacles

$$f(x,y) = a(\text{where a is a constant})$$

and differentiating it, we obtain

$$\left(\frac{\partial f}{\partial x}\right)dx + \left(\frac{\partial f}{\partial y}\right)dy = 0$$

From the equations

$$\left(\frac{\partial I}{\partial x}\right)dx + \left(\frac{\partial I}{\partial y}\right)dy = 0$$

$$\left(\frac{\partial f}{\partial x}\right)dx+\left(\frac{\partial f}{\partial y}\right)dy=0$$

we obtain

$$\left(\frac{\partial I}{\partial y}\right)\left(\frac{\partial f}{\partial x}\right)-\left(\frac{\partial I}{\partial x}\right)\left(\frac{\partial f}{\partial y}\right)=0$$

which, in combination with the equation of the obstacles, identifies the coordinates of the point of tangency.[66]

At this juncture Pareto pointed out that in the new formulation, the problem of economic equilibrium emerges only from "real experience, viz.: firstly, the individual's order of preference; secondly, the obstacles he encounters in these choices".[67] Hence, economic theory can finally "study the (economic) facts directly" and not through "the notions that men are possessed of".[68]

Pareto then specified that "if p_1, p_2, ... are the prices of the goods, q the cost of labour, i the interest rate, r, ... the price for lease of the land, etc. and if a_1, a_2, a_3 represent certain parameters, mathematical economics shows that the variables p_1, p_2, ..., q, i, r, are determined by all the parameters". In other words, given the system of equations

$$ax+by=c$$
$$a'x+b'y=c'$$

it makes no sense to ask "which of the parameters a, b, c, a', b', c' determines the value of x and y".[69]

Furthermore, in general equilibrium expressed as a system of equations, "the prices disappear by elimination" too, so that "for the determination of the quantities received by each person only the parameters of tastes, obstacles and wealth distribution remain".[70]

Pareto expounded his new conception of pure economics publicly for the first time in a short course he gave at the *École des Hautes Études Sociales* in Paris from the 10th to the 18th of November 1901. Here he began by specifying that economic equilibrium requires only the equations representing

the preferences and the balance of income and expenditure for each individual;[71]
the hypothetical social system (where the combination between private property and free competition is described by the equivalence between "the selling price of the products and their cost of production"; that between private property and monopoly is described by the condition that "the difference between the selling price [and] the cost of production is at a maximum"; collectivism is represented by production being performed in such a way as to "procure the greatest possible well-being for the citizens of the socialist State");[72]
"the relationships (not only technical [but] also economic) between the quantities of goods being transformed and the products derived from them".[73]

Pareto added that

after Irving Fisher, the conditions of capitalisation (relating to the production of those elements "which do not constitute the direct goal of production but a means of production") can "be based exclusively on the notion of the transformation of economic goods", thus allowing the notion of capital "[which is] arbitrary and not very rigorous, scientifically speaking"[74] to be dispensed with;
it is possible to limit ourselves to addressing the equilibrium of exchange by substituting the equations for production with those for exchange, since the latter encapsulate the fundamental idea that "what one individual finds himself short of, another acquires a surplus of".[75]

In the same period Pareto formalised these ideas in describing "any given state of equilibrium" characterise, for θ individuals, by the equations for maximum ophelimity (for commodities m, these are $\theta\ (m - 1)$), by the θ equations of equivalence between inflows and outflows and by m equations, one for each commodity, which "serve to indicate the obstacles which individuals will encounter in order to procure the economic goods for themselves", particularly the equivalence between quantities sold and purchased.[76] Thus, we have a system of $\theta m + m - 1$ equations (given that one of these depends on all the others) which allow us to identify the θm quantities exchanged and the $m - 1$ prices (one of the commodities is considered as the *numéraire* and hence a unit of this by definition constitutes the price).[77]

1.3 The *Manual of Political Economy*: The Innovations in Relation to Pareto's Previous Economic Ideas

On the 19th of November 1899, Pareto informed Pantaleoni that following up "the idea already touched on" in *Expounding the pure economics*, he was writing "a treatise on mathematical economics" in which he formulated "the fundamental equations without making use of either the final degree of utility, or ophelimity or even prices".[78]

This project, which he had already conceived a decade earlier,[79] took shape in the *Manual of Political Economy*[80] where he pointed out, firstly, that if the quantities of all the goods available to the individual increase (or decrease), "there is no problem to be solved" since evidently "the new position will be more (or less) advantageous for the individual involved".[81] Instead, economic problems consist precisely in ascertaining whether, following an increase in certain quantities and a decrease in others, "the new combination is or is not advantageous for the individual".[82]

In general, among the combinations of quantities of goods available, the individual's choice can be established by reference to "the theory of economic equilibrium",[83] hinting, in conceptual terms, at the result of the "contrast between people's preferences and the obstacles to satisfying them"[84] and in formal terms, to "the state … [in which] no further exchanges will occur"[85] since "the exchanges permitted by the obstacles are prevented by the preferences [and] vice versa".[86]

Economic problems are particularly manifested in exchanges (involving "giving one thing to receive another") and in production (where "certain things are transformed into certain other things").[87]

Further, an individual may engage in the exchange "at [prevailing] market conditions" or may modify these.[88] Hypothetically, in the first case (corresponding to "free competition"),[89] the individual aims simply to "satisfy his own desires"[90] whereas in the second case (corresponding to "monopoly"),[91] he seeks the attainment of the market conditions which will allow him to achieve the end he was aiming for,[92] with the implication that under equilibrium, the quantities of goods corresponding to each case are different.[93] Similarly, an enterprise may accept the prevailing prices or may modify them,[94] but in either case with the objective of obtaining "the maximum cash monetary profit … [by] pay[ing] the lowest amount possible for its purchases and obtain[ing] the highest amount possible for what it sells".[95]

Tastes (Preferences)

Having established this, Pareto observed that if "everyone used the goods he possessed for only as long as he liked",[96] then pure economics should consider "not the quantities consumed, but the quantities available to the individual"[97] as well as "the present anticipation of the future consumption of the goods available, as constituting the motive for the individual's actions".[98]

Moreover, Pareto noted that "in general, consumption is dependent" in two ways.[99] The first of these (relating to "complementary goods") refers to "the pleasure of consumption being in relation to the pleasures of alternative consumption ... over a broad range of variation [in the quantities of goods]".[100] The second (relating to substitute goods) refers to "being able to substitute one thing for another so as to provide an individual with sensations which, if not identical, are at least approximately the same".[101] Lastly there also remains the case of independent goods, that is, that whereby, through "restricted variations in the quantities of the goods, the ophelimity deriving from the consumption of an item [is] independent [of] the consumption of the others".[102]

In the light of all this, an individual's preferences can be expressed through a series of infinite combinations of quantities of the same goods between which the individual "would be unable to choose".[103] For all three of the typologies of goods mentioned,[104] the series in question can be represented graphically by means of a curve of indifference[105] which can be imagined as continuous and which has the properties, recognised thanks to "everyday experience", $(dy/dx) < 0$,[106] $(d^2y/dx^2) > 0$,[107] that is, in graphic terms a negatively inclined curve which is strictly convex.

In algebraic terms, having defined x, y, z the quantities of commodities X, Y, Z ... available to the individual and b_y, c_z, ... the increases which X must undergo in order to compensate, in the eyes of the individual, for the decreases in the quantities of Y, Z, ...,[108] the curve of indifference can be constructed as follows:[109]

if dx represents the increase in X which compensates the reduction in overall ophelimity caused by the reduction dy in the quantity of Y available, then the equation

$$dx + b_y dy = 0$$

or[110]

$$\left(\frac{\partial x}{\partial y}\right)dy + b_y dy = 0$$

can apply, and likewise the equations

$$\left(\frac{\partial x}{\partial z}\right)dz + c_z dz = 0$$

$$\ldots$$

which, added together, yield the equation

$$\left(\frac{\partial x}{\partial y}\right)dy + b_y dy + \left(\frac{\partial x}{\partial z}\right)dz + c_z dz + \ldots = 0$$

which, since

$$dx = \left(\frac{\partial x}{\partial y}\right)dy + \left(\frac{\partial x}{\partial z}\right)dz + \ldots$$

leads to[111]

$$dx + b_y dy + c_z dz + \ldots = 0$$

Pareto strongly underlined that this equation, where "the quantities b_y, c_z, ... are the product of experience",[112] "is the only one which we need, strictly speaking, to formulate the theory of economic equilibrium".[113]

If

total ophelimity does not depend on the order of consumption and the ophelimity resulting from the consumption of a given quantity of a commodity depends only on that consumption;

total ophelimity depends on the order of consumption, of which we have in any case an empirical knowledge;[114]

the above equation can be integrated, thus allowing us to arrive at the equation of a curve of indifference:[115]

$$I = \Phi(x,y,z,\ldots)$$

where I is the index.[116]

By differentiating

$$I = \Phi(x,y)$$

we obtain

$$0 = \varphi_x dx + \varphi_y dy$$

from which (since dx and dy are of opposite signs) we can conclude that the elementary ophelimities $\varphi_x(=\partial I/\partial x)$ e $\varphi_y(=\partial I/\partial y)$ can both have a positive sign,[117] which constitutes the first algebraic property of the indices of indifference.[118]

Further, from

$$0 = \varphi_x dx + \varphi_y dy$$

we can obtain

$$-\left(\frac{\varphi_x}{\varphi_y}\right) = \frac{dy}{dx}$$

from which we arrive at

$$\left[-\frac{\partial\left(\frac{\varphi_x}{\varphi_y}\right)}{\partial x}\right] > 0$$

$$\left[-\frac{\partial\left(\frac{\varphi_x}{\varphi_y}\right)}{\partial y}\right] < 0$$

developing which we obtain

$$\left[-\frac{\partial\left(\frac{\varphi_x}{\varphi_y}\right)}{\partial x}\right] = \left(\varphi_{xx}\varphi_y - \varphi_{xy}\varphi_x\right) < 0$$

$$\left[-\frac{\partial\left(\frac{\varphi_x}{\varphi_y}\right)}{\partial y}\right] = \left(\varphi_{yy}\varphi_x - \varphi_{xy}\varphi_y\right) < 0$$

which, supposing $\varphi_{xy} = 0$[119] and remembering that $\varphi_x > 0$ e $\varphi_y > 0$, imply

$$\varphi_{xx} < 0 \text{ and } \varphi_{yy} < 0$$

which represent the second property of the indices of indifference, that is, the decrease in the elementary ophelimity of a commodity with increase in quantities available of that commodity.[120]

The Obstacles

The obstacles contributing to the determination of general equilibrium are of two types.

The first type[121] comprises:

"the preferences of those with whom the individual is dealing";
the fact that "the quantity of goods to be divided between various individuals is a given";

the fact that a commodity "can be procured [only] by using other commodities": Pareto viewed production as the means adopted to overcome this type of obstacle;[122]
"obstacles deriving from the social order".

The second type, instead, consists of anything that limits the route followed by the individual in order to pass from the combination of quantities of goods initially available to his preferred combination.[123] Pareto defined the route as the set of combinations of quantities effectively available to the individual, representing it with a continuous negatively inclined curve,[124] not necessarily a straight line even if the "straight routes are more common in practice".[125]

The type of obstacle on which Pareto dwelt most was production, whose "elements" he divided into those which "are not consumed or are consumed slowly", alternatively referred to as "capital"[126] and those which "are consumed rapidly",[127] alternatively referred to as "consumables".[128] He then underlined that among the various means available for producing a commodity, the producer will choose on the basis of "not only technical, but also economic considerations".[129]

Generalising, Pareto specified that one commodity can be transformed into another (i.e. within a production process) either materially or by transporting it either in time or in space.[130]

Material transformation is represented graphically by a continuous, positively inclined curve:[131] with increases in the quantities used of the commodity to be transformed, the quantity of the commodity obtained from the transformation can increase more than proportionally (in which case the curve is always strictly convex[132]) or less than proportionally (in which case the curve is always strictly concave).[133] This curve is denominated a zero-index curve of indifference to obstacles or curve of complete transformations: it shows the quantity of a commodity that can be produced from any quantity of another commodity, "without leaving any residue".[134] If this curve is shifted to the right (left) of a unitary segment, a curve of indifference to obstacles with an index of plus (minus) one is generated, because in the case of a plus one index, once a given quantity of a commodity has been produced, a unitary segment of the starting commodity is still available; if the index is minus one, on the other hand, an additional unitary segment of the starting commodity is required to produce a given quantity of the end-commodity.[135] If there is only one producer, these lines of indifference will also be lines of iso-profit, positive (negative) if the index is positive (negative).[136]

Individual Equilibrium

The point of tangency between a route and a curve of indifference represents the highest point on the route followed by an individual,[137] while any other point where the individual is obliged to stop by the obstacles is referred to as the terminal point of the route.[138] The curve with which these points can be joined[139] is termed "line of equilibrium of the preferences [or] line of exchange"[140] and "can also be called the curve of supply and demand" from the consumer's point of view,[141] where in the "case of more than one commodity the supply of a given commodity depends on the prices of all the other commodities exchanged and likewise the demand for a commodity".[142]

Pareto[143] observed that "adding up the quantities of commodities transformed by each individual" yields the curve of the exchanges of the group of individuals: along this line, "the members of the group in question derive the maximum of ophelimity",[144] with this latter described again as that "position any minute deviation from [which has] necessarily the effect of benefitting certain members of the group and harming others".[145]

In algebraic terms, if we suppose that the route followed by the individual is represented by the equation[146]

$$f\left(x,y,z,\ldots.\right)=0$$

then by deriving it, if the number of commodities X, Y, Z ... is m, the m - 1 equations

$$f_x\left(\frac{\partial x}{\partial y}\right)+f_y=0; f_x\left(\frac{\partial x}{\partial z}\right)+f_{yz}=0;\ldots$$

or the equal number of equations

$$\left(\frac{\partial x}{\partial y}\right)=-\frac{f_y}{f_x};\left(\frac{\partial x}{\partial z}\right)=-\frac{f_z}{f_x};\ldots$$

can be obtained.

If we have the equation for the curve of indifference

$$I = (x, y, z, \ldots)$$

then the m - 1 equations (corresponding to first degree conditions for the maximisation of the function)

$$\varphi_x \left(\frac{\partial x}{\partial y} \right) + \varphi_y = 0; \varphi_x \left(\frac{\partial x}{\partial z} \right) + \varphi_z = 0; \ldots$$

or the equal number of equations

$$\frac{\partial x}{\partial y} = -\frac{\varphi_y}{\varphi_x}; \frac{\partial x}{\partial z} = -\frac{\varphi_z}{\varphi_x}; \ldots$$

can be obtained.

By combining these two series of equations, we obtain the m - 1 equations[147]

$$\frac{f_y}{f_x} = \frac{\varphi_y}{\varphi_x}; \frac{f_z}{f_x} = \frac{\varphi_z}{\varphi_x}; \ldots$$

or the equal number of equations

$$\varphi_x = \left(\frac{f_x}{f_y} \right) \varphi_y = \left(\frac{f_x}{f_z} \right) \varphi_z; \ldots$$

which, together with the route equation, constitute the system of m equations whose solution yields the quantities of commodities X, Υ, Z, ... which maximise the individual's ophelimity.

Pareto also reformulated the argument incorporating the prices: again, following

$$\frac{\partial x}{\partial y} = -\frac{f_y}{f_x} \text{ or } -\frac{\partial x}{\partial y} = \frac{f_y}{f_x}$$

and the corresponding equations

$$-\frac{\partial x}{\partial y} = \frac{f_y}{f_x} \ldots$$

the ratios f_y/f_x; f_z/f_x … can be defined as the prices of Y, Z… expressed in terms of X.

These prices, which are genuinely observable entities, are therefore formulated as[148]

$$\frac{f_y}{f_x} = p_y; \frac{f_z}{f_x} = p_z \ldots.$$

and it is the integration of these which leads us to the route equation mentioned above[149]

$$f(x,y,z,\ldots) = 0$$

According to Pareto,[150] "in a very large number [of] highly significant phenomena, the prices can be considered as constant" so that the integration of the equations

$$\frac{\partial x}{\partial y} = -p_y; \frac{\partial x}{\partial z} = -p_z \ldots$$

gives rise to the following specific form of the route function[151]

$$x + p_y y + p_z z + \ldots = c$$

and also to[152]

$$x_0 + p_y y_0 + p_z z_0 + \ldots = c$$

By subtracting the second equation from the first we obtain

$$(x - x_0) + p_y(y - y_0) + p_z(z - z_0) + \ldots = 0$$
$$\text{or } dx + p_y dy + p_z dz + \ldots = 0$$

which "yields the balance of income and expenditure for the individual concerned".

On this basis, Pareto[153] pointed out that given an individual, m commodities and constant prices, the equilibrium emerges from the balancing equation

$$(x - x_0) + p_y(y - y_0) + p_z(z - z_0) + \ldots = 0$$

in association with the m - 1 equations for the maximisation of the individual's ophelimity

$$\varphi_x = \left(\frac{f_x}{f_y}\right)\varphi_y = \left(\frac{f_x}{f_z}\right)\varphi_z; \ldots$$

which, bearing in mind the equations

$$\frac{f_y}{f_x} = p_y; \frac{f_z}{f_x} = p_z \ldots.$$

can also be formulated as[154]

$$\varphi_x = \frac{\varphi_y}{p_y} = \frac{\varphi_z}{p_z} \ldots$$

or as[155]

$$p_y = \frac{\varphi_y}{\varphi_x}; p_z = \frac{\varphi_z}{\varphi_x}; \ldots$$

And, in fact, the result is a system of m equations whose solution yields the quantities of commodities X, Y, Z, … which maximise the ophelimity of the individual. On the basis of this system, Pareto, following a much more complicated[156] procedure than in his original attempt of 1892,[157] determined the laws of demand and of supply for a commodity in conditions of general equilibrium, confirming, in particular, that with increases in the price of a commodity, demand will diminish, while nothing can be concluded with regard to the sign of the variation in supply.[158]

Lastly, Pareto hypothesised that the individual,[159] starting with the quantity x_0 of commodity X: consumes quantity a, transforms quantity by into Y (where b represents the number of units of X required to obtain a unit of Y) and ends with quantity x definitively at his disposal. This hypothesis is expressed through the equation

$$x_0 = a + by + x$$

or alternatively

$$a + by + x - x_0 = 0$$

deriving which in relation to x

$$b\left(\frac{\partial y}{\partial x}\right) + 1 = 0$$

or, alternatively

$$\left(\frac{\partial x}{\partial y}\right) = -b$$

substituting which in

$$\varphi_x\left(\frac{\partial x}{\partial y}\right) + \varphi_y = 0$$

yields

$$\varphi_y - b\varphi_x = 0$$

which, in combination with the route equation specified earlier, allows us to determine the system of two equations in the variables x and y whose solution yields the quantities of X and Y available following the transformation and hence also the quantity of X, equal to $x_0 - x$, which has been transformed into Y.

In relation to a given route, the producer, for his part, can follow it: to a point of tangency with a curve of iso-profit, with this point representing its point of equilibrium and with competition being, in Pareto's opinion, "incomplete".[160] Instead, Pareto considers competition to be "complete" if the route has no point of tangency or if the point of tangency corresponds to a "[profit] index lower than the neighbouring index points on the route".[161]

The General Equilibrium of Exchange

Pareto's graphic representation of the exchange between two individuals consisted of

a rectangle, constituting the *locus* of the points of the plane whose coordinates indicate the infinite ways in which given quantities of two commodities (the dimensions of the rectangle) can be divided between two individuals (this figure was later to acquire renown as the Edgeworth box);[162]
the families of the lines of indifference of the two individuals (or traders), placed within the rectangle and originating from its south-western and north-eastern vertices;
a common route for the two individuals.

If this common route is a straight line, there will be a general point of equilibrium corresponding to the point where the route is tangential to the curves of indifference of both individuals: this general point of equilibrium also corresponds to the point of intersection between the lines of exchange of the two individuals mentioned.[163] The point of equilibrium, where the two individuals are "conducted by competition",[164] may be stable (i.e. such that "if the two individuals deviate [from it], they then tend to return to it"[165]) or unstable (i.e. such that if the two individuals deviate from it, they diverge ever further away[166]).

If, on the other hand, the common route is a curve, it will touch the two individuals' lines of indifference at two different points so that there is no general equilibrium unless one of the two is able to impose on the other his own preferred route,[167] which conducts him to his own point of maximum ophelimity,[168] corresponding to the point of tangency between the other individual's line of exchanges and his own curve of indifference.[169]

In algebraic terms,[170] let *X* and *Y* denote the commodities exchanged; 1 and 2 the two individuals; x_{10}, y_{10}, x_{20}, y_{20} the quantities of the two commodities initially held by the two individuals and x_1, y_1, x_2, y_2 the quantities available to them at equilibrium.

As "the total quantities of each commodity remain constant", the following equations apply to the exchange

$$x_{10} + x_{20} = x_1 + x_2 ; y_{10} + y_{20} = y_1 + y_2$$

If both individuals follow their own routes, which hypothetically coincide with each other, "without worrying about anything other than arriving at a point of equilibrium [, as is] the case in the context of free competition",[171] then for individual 1 the following equations[172]

$$f(x_1, y_1, \mu) = 0,$$

$$\varphi_{1y} f_{1x} - \varphi_{1x} f_{1y} = 0$$

will apply, from which the equation of the route of individual 2 can be derived

$$f(x_{10} + x_{20} - x_2, y_{10} + y_{20} - y_2, \mu) = 0$$

Individual 2 will follow this route to the point where he maximises his own ophelimity, that is, the point in which[173]

$$\varphi_{2x} dx_2 + \varphi_{2y} dy_2 = 0 \, or \, \varphi_{2x} dx_1 + \varphi_{2y} dy_1 = 0$$

since

$$d(x_{10}+x_{20})=d(x_1+x_2)\text{ gives rise to }0=(dx_1+dx_2)\text{ or to }dx_2=-dx_1$$

$$d(y_{10}+y_{20})=d(y_1+y_2)\text{ gives rise to }0=(dy_1+dy_2)\text{ or to }dy_2=-dy_1$$

$$\text{As }\frac{dx}{dy}=-\frac{f_y}{f_x}\text{ or }\frac{dx_1}{dy_1}=-\frac{f_{1y}}{f_{1x}}$$

$$\varphi_{2x}dx_1+\varphi_{2y}dy_1=0$$

will be transformed, by substitution, into

$$\frac{\varphi_{2y}}{\varphi_{2x}}=-\left(\frac{dx_1}{dy_1}\right)=\frac{f_{1y}}{f_{1x}}$$

or, alternatively,

$$\varphi_{2y}f_{1x}-\varphi_{2x}f_{1y}=0$$

Thus, we obtain a system of five independent equations

$$x_{10}+x_{20}=x_1+x_2;\,y_{10}+y_{20}=y_1+y_2;\,f(x_1,y_1,\mu)=0;$$
$$\varphi_{1y}f_{1x}-\varphi_{1x}f_{1y}=0;\varphi_{2y}f_{1x}-\varphi_{2x}f_{1y}=0$$

resolvable into the five variables x_1, y_1, x_2, y_2, μ.

Once again referring to a regime of free competition, Pareto's treatment of the general equilibrium of exchange in relation to θ individuals 1, 2, … and m commodities X, Y, Z, …, with the same prices for all the individuals[174] and, for convenience, considered as constant,[175] was developed as follows: He initially maintained that in the exchange "overall quantities remain constant",[176] so that the following m equations apply[177]

$$x_1-x_{10}+x_2-x_{20}\ldots=0=X-X_0;y_1-y_{10}+y_2-y_{20}\ldots=0=Y-Y_0;\ldots$$

Subsequently, Pareto remarked that the route of each individual should concern only "the quantities which relate exclusively to that individual".[178] Consequently, there will be θ routes having the following general form, for example in the case of individual 1:

$$f_1\left(x_1, y_1, z_1, \dots\right) = 0$$

The prices of *Y*, *Z*, ... in terms of *X*, for individual 1, can be expressed in the form[179]

$$p_y = -\frac{\partial x_1}{\partial y_1} = \frac{f_{1y}}{f_{1x}}; p_y = -\frac{\partial x_1}{\partial z_1} = \frac{f_{1z}}{f_{1x}}; \dots$$

and since we postulate that they are the same for all the individuals, we can also obtain the equations

$$p_y = \frac{f_{1y}}{f_{1x}} = \frac{f_{2y}}{f_{2x}} = \dots; p_z = \frac{f_{1z}}{f_{1x}} = \frac{f_{2z}}{f_{2x}} = \dots; \dots$$

It is subsequently possible to postulate that the θ equations of the individual routes referred to above can be specified as follows:[180]

$$f_1 = x_1 - x_{10} - p_y\left(y_1 - y_{10}\right) + \dots = 0; f_2 = x_2 - x_{20} - p_y\left(y_2 - y_{20}\right) + \dots = 0; \dots$$

Pareto then showed that the equation $X - X_0 = 0$, depending as it does on the other $m - 1$ equations of the quantities exchanged and on the θ equations of the individual routes, "must be discarded",[181] with the corollary that the equations of the quantities exchanged are reduced to $m - 1$.

The system of equations for the general equilibrium of exchange is completed with the familiar $(m - 1)$ equations for the maximisation of individual ophelimity:

$$\varphi_{1x} = \frac{\varphi_{1y}}{p_y} = \dots; \varphi_{2x} = \frac{\varphi_{2y}}{p_y} = \dots; \dots$$

which, however, apply only for a situation of equilibrium.[182]

The system is therefore made up of the $m - 1$ equations of quantities exchanged, the θ specific equations of the individual routes and the $(m - 1)$ equations for the maximisation of individual ophelimities, that is, by $\theta m + m - 1$ independent equations which allow the determination of the

θm quantities available to the individuals at the termination of the exchange and the m - 1 prices[183] (with commodity x treated as the *numéraire*).

Pareto also studied cases of general equilibrium in monopoly exchanges.[184] If for the monopolist (e.g. trader 1) the commodity monopolised has a null ophelimity,[185] then the equations of his maximum ophelimity, and hence those of the system of which they form part, are reduced by one;[186] to preserve the resolvability of the system, it is thus necessary also to reduce the number of variables by one. To this end the monopolist may fix the price of the commodity, for example, by adopting the price which maximises the sales revenue or his ophelimity.[187]

Algebraically, if

the commodity monopolised is Y; the quantity of it sold is $y_{10} - y_1$, where y_{10} represents the initial quantity and y_1 the remaining final quantity available; p_y represents the price of the commodity;

the equation for the revenue is

$$p_y(y_{10} - y_1) = p_y f(p_y)$$

which is maximised by the value of p_y, for which

$$\frac{d\left(p_y f(p_y)\right)}{dp_y} = 0$$

Alternatively, if the ophelimity function of 1 is

$$\varphi_1(x_1, y_1, z_1, \ldots)$$

the monopolist maximises it by imposing the value of p_y for which

$$\left[\frac{d\varphi_1(x_1, y_1, z_1, \ldots)}{dp_y}\right] = 0$$

in other words, for which

$$\varphi_{1x}\left(\frac{dx_1}{dp_1}\right)+\varphi_{1y}\left(\frac{dy_1}{dp_1}\right)+\varphi_{1z}\left(\frac{dz_1}{dp_1}\right)+\ldots=0$$

In the same way, if commodity *Y* has two monopolists[188] (e.g. traders 1 and 2), the equations maximising the ophelimity of the monopolists are reduced by two and hence those determining the system of general equilibrium are reduced by the same quantity: in order to ensure the resolvability of these, the number of variables must also be reduced by two, for example, by fixing p_y and y_2 (i.e. the quantity of *Y* which remains available to 2 after the exchange).[189] In these cases, Pareto showed that it is impossible for both monopolists to maximise their revenue or their ophelimity,[190] at the same time pointing out that there are "very numerous and very diverse cases" where either the two monopolists are reduced to one (corresponding to cartels) or one of the monopolists accepts the price fixed by the other, normally the one who controls the greater proportion of production of the commodity.[191]

Lastly, let 1 represent the monopolist of commodity *Y* and 2 the monopolist of commodity *Z*. In such a context, since *Y* does not contribute to the ophelimity of 1 nor *Z* to that of 2, the equations for the maximisation of the monopolists' ophelimity will be reduced by two: in order to preserve the resolvability of the system of general equilibrium the number of variables must also be reduced, for example, by fixing p_y e p_z.[192] In these circumstances, Pareto showed that both monopolists can maximise their revenue and their ophelimity.[193]

The General Equilibrium of Production and of Exchange

From a formal perspective,[194] given θ individuals, production refers to the transformation of n commodities (or capital services) A, B, ... whose prices are p_a,[195] p_b, p_c, ... into the m commodities X, Y, ..., whose cost prices are π_x, π_y,... and whose sales prices are p_x, p_y, ... For the sake of simplicity, only the initial quantities and the equilibrium quantities of the commodities are taken into account.

On this basis, for capital services we have the following definitions:

The initial quantities are

$$a_{10}+a_{20}+\ldots=A_0;b_{10}+b_{20}+\ldots=B_0;\ldots$$

the quantities consumed by the enterprises in equilibrium are

$$a_1^{'} + a_2^{'} + \ldots = A'; b_1^{'} + b_2^{'} + \ldots = B'; \ldots$$

and hence the equilibrium quantities contributed by the consumers to the enterprises are

$$A_0 - A' = A''; B_0 - B' = B''; \ldots$$

while the quantities transformed by the enterprises in equilibrium are $A''', B''', \ldots$

For the commodities produced, whose initial quantities are hypothesised as being nil, we have the following definitions:

the quantities consumed by the consumers in equilibrium are

$$x_1^{'} + x_2^{'} + \ldots = X'; y_1^{'} + y_2^{'} + \ldots = Y'; \ldots$$

while the quantities produced by the enterprises in equilibrium are $X'', Y'' \ldots$

Pareto specified that conditions of general equilibrium[196] in a free market context will result in the following:[197]

$$X' = X''; Y' = Y''; \ldots$$

Pareto then underlined that "the task of the company" is to determine the production coefficients (the minimum quantity of a commodity required to produce a unit of another commodity), bearing in mind that these can vary not only with variations in the quantities produced but also "in such a manner that their increases can be compensated by reductions elsewhere".[198]

In order to determine the production coefficients, the "technical parameters" of production are represented by the functions[199]

$$A = F(x, y, \ldots); B = G(x, y, \ldots); \ldots$$

which yield the production coefficients

$$a_x = \frac{\partial F}{\partial x}; b_x = \frac{\partial G}{\partial x}; \ldots$$

$$a_y = \frac{\partial F}{\partial y}; b_x = \frac{\partial G}{\partial y}; \ldots$$

$$\ldots$$

Supposing that the production coefficients are constant, that X, Υ, ... are produced independently and that the prices of transformed commodities are constant, the costs of production for small increments of X, Υ, ... are[200]

$$(a_x + p_b b_x + \ldots) dx = \pi_x dx$$

$$(a_y + p_b b_y + \ldots) dy = \pi_y dy$$

$$\ldots$$

On this basis, the general equilibrium of production and of exchange is constituted:[201]

of the $(m + n - 1)\theta$ equations of the weighted ophelimities of commodities A, B, ... X, Υ, ...

$$\varphi_{1x} = \frac{\varphi_{1y}}{p_y} = \ldots = \varphi_{1a} = \ldots;$$

$$\varphi_{2x} = \frac{\varphi_{2y}}{p_y} = \ldots = \varphi_{2a} = \ldots;$$

$$\ldots$$

of the θ equations of the individual routes

$$a_1' - a_{10} + p_b (b_1' - b_{10}) + \ldots p_x x_1' + p_y y_1' \ldots = 0$$

$$a_2' - a_{20} + p_b (b_2' - b_{20}) + \ldots p_x x_2' + p_y y_2' \ldots = 0$$

$$\ldots$$

of the m equations which represent "the parity of the cost of production and the selling price" for each unit of a commodity produced

$$p_x = a_x + p_b b_x + \dots$$
$$p_y = a_y + p_b b_y + \dots$$
$$\dots$$

of the n equations indicating that for each commodity supplied to the enterprises, the quantity supplied is fully utilised in the production of other commodities, represented by the quantities consumed in equilibrium, which under conditions of free competition are equal to the quantities produced

$$A_0 - A' = a_x X' + a_y Y' + \dots$$
$$B_0 - B' = b_x X' + b_y Y' + \dots$$
$$\dots$$

Multiplying each of these last equations by the price of the commodity they refer to and summing them will yield the equation

$$A_0 - A' + p_b \left(B_0 - B'\right) + \dots = a_x X' + a_y Y' + \dots p_b b_x X' + p_b b_y Y' + \dots$$

that is,

$$A_0 - A' + p_b \left(B_0 - B'\right) + \dots = \left(a_x + p_b b_x + \dots\right) X' + \left(a_y + p_b b_y + \dots\right) Y' + \dots$$

or

$$A_0 - A' + p_b \left(B_0 - B'\right) + \dots = p_x X' + p_y Y' + \dots$$

or again

$$A' - A_0 + p_b \left(B' - B_0\right) + \dots + p_x X' + p_y Y' + \dots = 0$$

the same equation that we can obtain by summing

$$a'_1 - a_{10} + p_b\left(b'_1 - b_{10}\right) + \ldots p_x x'_1 + p_y y'_1 \ldots = 0$$

$$a'_2 - a_{20} + p_b\left(b'_2 - b_{20}\right) + \ldots p_x x'_2 + p_y y'_2 \ldots = 0$$

We are therefore faced with a system made up only of $(m + n)\theta + m + n - 1$ independent equations which is nevertheless resolvable, since, with these equations, there are an equal number of variables represented by the $m + n - 1$ prices p_x, p_y, ... p_b, p_c, ... and by the $(m + n)\theta$ quantities x'_1,... a'_1...[202].

In graphic terms, assuming producers and consumers do not aim to modify market conditions, their collective representation, starting from the same route, consists, for the former, in the curve of complete transformations (of maximum earnings), provided they are acting in the context of a regime of complete (or incomplete) competition, while, for the latter, it consists in the line of exchanges.[203] The points of general equilibrium are those located either on the intersection between the line of exchanges and the curve of complete transformations[204] (in the case of complete competition) or between the line of exchanges and the line of maximum revenue (in the case of incomplete competition).[205]

If the producer is able to influence the market, he will seek to "penetrate as far as possible into the region of the positive indices", with the implication that general equilibrium will be imposed by him as the point of intersection between the line of exchanges and one of the producer's curves of indifference "in the case of complete competition" or as the point of tangency between the line of exchanges and the line of maximum revenue "in the case of incomplete competition".[206]

If, finally, the producer coincides with the consumer, he will select the point on the curve of complete transformations which is tangential to a curve of his preferences.[207]

Maximum Ophelimity of a Collectivity

Some time after this, Pareto, commenting on an early version of an article wherein the young mathematician and statistician from Trieste Luigi Vladimiro Furlan sought to apply the concept of ophelimity to a collective context,[208] stated that Furlan's attempt, if he truly wished to "advance a general theory", would need to refer to a group of individuals who were "absolutely similar or supposed such ... so that in reality it would always be better to speak about a single individual" and that it was only at the applied level that it would be possible to "try, by approximation, to form groups".[209]

Thus, in April 1913, Pareto[210] observed that "sociology needs to find a means", distinct from that used in political economy,[211] to render the variation in the ophelimities of single individuals homogeneous. Any given individual (e.g. [1]) can set out to "operate in a manner such that all his fellow citizens attain the greatest possible level of well-being without anyone being sacrificed". For this reason the individual in question, who has "direct experience" only of $\delta\varphi_1$, "must imagine the variations $\delta\varphi_2$, $\delta\varphi_3$, ..." referred to other individuals, which he thus transforms from "objective and heterogeneous" to "subjective and heterogeneous" by multiplying them by the coefficients α_{12}, α_{13}, ... (where α_{12} represents the assessment given by 1 concerning the variations in the ophelimity of 2, etc.). Thus, there exist as many different conceptions of the maximum of collective ophelimity as there are individuals and each of these possible conceptions can be expressed through a corresponding number of equations, with the coefficients of these different equations being heterogeneous among themselves

$$
\begin{aligned}
0 &= \alpha_{11}\delta\varphi_1 + \alpha_{12}\delta\varphi_2 + \alpha_{13}\delta\varphi_3 + \ldots \\
0 &= \alpha_{21}\delta\varphi_1 + \alpha_{22}\delta\varphi_2 + \alpha_{23}\delta\varphi_3 + \ldots \\
0 &= \alpha_{31}\delta\varphi_1 + \alpha_{32}\delta\varphi_2 + \alpha_{33}\delta\varphi_3 + \ldots \\
&\ldots
\end{aligned}
$$

In order to render homogeneous the coefficients which the various individuals have attributed to $\delta\varphi_1$, $\delta\varphi_2$, $\delta\varphi_3$, ... they must be multiplied by new coefficients β_1, β_2, β_3, ... determined by someone else, such as the government, "to an objective end, such as collective prosperity" (so that β_1 represents the importance assigned by the government to the assessments made by 1, etc.). In this fashion the equations

$$
\begin{aligned}
M_1 &= \alpha_{11}\beta_1 + \alpha_{21}\beta_2 + \alpha_{31}\beta_3 + \ldots \\
M_2 &= \alpha_{12}\beta_1 + \alpha_{22}\beta_2 + \alpha_{32}\beta_3 + \ldots \\
M_3 &= \alpha_{13}\beta_1 + \alpha_{23}\beta_2 + \alpha_{33}\beta_3 + \ldots \\
&\ldots
\end{aligned}
$$

are obtained, with M_1 representing the importance which the group attributes to the variation in 1's ophelimity, via the homogenising assessment of the government, and so on.

Thus, it will be possible to sum the foregoing equations and arrive at the new equation

$$0 = M_1 \delta \varphi_1 + M_2 \delta \varphi_2 + M_3 \delta \varphi_3 + \ldots$$

The government, in halting the collective movement at this maximum point, will avoid "inflicting unnecessary suffering on the whole group or on a part of it".

Various maxima can exist in sociology. For exemple, the maximum quantity of the population conforming to the maximum of collective utility is greater (for reasons of "military and political power") than the maximum quantity of the population conforming to the maximum of collective utility established taking into account, for each social class, not only of the political and military benefits but also of the economic costs of each demographic increment.

1.4 The Definitive Formulation of Pareto's Conception of General Economic Equilibrium

In March 1912,[212] in an article he prepared for a French mathematical encyclopaedia, Pareto expressed his hope that "[he had] succeeded in presenting the theory of pure economics in the clearest manner yet".

In this article, having defined the prices in terms of subjective ratios of exchange between differing quantities of commodities, that is, (given that x represents the *numéraire* and thus $p_x = 1$), as

$$p_y = -\frac{\partial x}{\partial y}; p_z = -\frac{\partial x}{\partial z}; \ldots$$

Pareto claimed that "in order to approximate reality, it is better to select the variables in such a manner that prices are always positive".[213]

Let us therefore imagine two commodities, X and Y, the first of which is on offer (with an initial quantity of x_0) and the second in demand (with an initial quantity of nil) at a price of p. In geometric terms the exchange will conclude at a point "which can be determined by observation"; by modifying the straight line of the price, the point of termination of the exchange will be shifted, with the set of such points denominated the "line of supply and demand"[214] which, for the case of n commodities, can be formulated as[215]

$$F\left(x, y, \ldots, x_0, y_0, \ldots\right) = 0$$

Modifying the values of x_0 and of p will yield other similar lines with which it will be possible to cover the entire Cartesian plane: "After that the individual can disappear; he is no longer needed for the determination of economic equilibrium, the lines are sufficient".[216] Similar lines could be obtained if the individual were to move along a family of parallel curves, each passing through a combination of quantities of commodities, which can be generalised as index functions.[217] Pareto specified that "from a strictly mathematical point of view, it is not necessary to make reference to the functions of supply and demand or to index functions for the determination of equilibrium".[218]

The equilibrium of an individual, if he is free of constraints,[219] occurs when the differential of the chosen form of his index function is annulled, that is, when it reaches a maximum or a minimum point. However, "the indices referring to different individuals are heterogeneous quantities and there is no question of adding them together, nor of finding the maximum of a sum that does not exist".[220]

Having said all this, Pareto stated that political economy pertains to the case where the equations (or the constraints) are less numerous than the variables, thus allowing "certain movements which will be compatible with the liaisons". As for the index functions, "the index functions determine the points where, with these movements, a position of equilibrium is attained".[221]

The constraints to be examined are suggested by the observation of "concrete examples",[222] of which there are two types:

the first type consists of the budget equations for each single individual, together with the budget equations of the entire group of individuals and the equations indicating whether the quantity of commodities, in passing from the initial state to that of equilibrium, is or is not completely transformed into additional quantities of other commodities;[223]

the second type consists of the equations describing the route leading from the initial quantities of the commodities to the equilibrium quantities.[224] These equations concern either the individuals or the transformation of the commodities, with the quantity transformed of each commodity being equal to the sum of the increments to the production to which it contributes multiplied by the manufacturing coefficients.[225]

In general, Pareto distinguished between economic phenomena on the basis of whether, in a state of equilibrium, the parameters of the equations are considered constant with variations in a variable (first type) or are similarly variable (second type);[226] in the first type the individuals "accept the market prices as they stand", while in the second type "they know how to, wish and are able to modify these prices" either in their own interest or in that of the group.[227]

In relation to production

the first type is characterised by the correspondence between "the cost of production [and] the selling price, [not] only as regards the total quantity but also for the last batch produced";

the second type, if performed in the interest of the individuals who adopt it, is characterised by the fact that "the selling price is higher than the cost of production"; if, on the other hand, it is performed in the interest of the group, "no relation exists, a priori, between the cost of production and the selling price", since this relation depends on the precise interpretation given to the maximum of the group which is being pursued.[228] Pareto then acknowledged that once "the rules of distribution deemed appropriate had been fixed", the maximum ophelimity for a group corresponds to that "position" a departure from which "in the smallest degree" will lead to an increase in well-being for some members and a decrease for others.[229]

Having specified this, Pareto likewise confirmed that given $\theta + 1$ individuals and m commodities, a system can be constructed consisting of:[230]

$(m - 1)(\theta + 1)$ equations representing the equality, for each individual, of the weighted elementary ophelimities with their prices, thus suggesting that "given the liaisons, the individuals' preferences are satisfied";

m equations establishing that the summation of the quantities available to the individuals following the exchange is equal to the summation of the quantities available to the individuals prior to the exchange, i.e. indicating that "the total quantities of commodities remain constant";

θ equations establishing the account balance of each individual whereby, for each, the value for the quantities of commodities at his disposal prior to the exchange is equal to the value for the quantities of commodities at his disposal after the exchange.[231]

We are thus left with $m\theta + 2m - 1$ equations featuring a corresponding number of variables, that is, $m - 1$ prices and $m(\theta + 1)$ quantities available to the individuals following the exchange, showing that the general equilibrium of exchanges under free competition "is well-founded".[232]

Again under free competition, in order to extend the general equilibrium of exchange to that of production and of exchange, it is necessary to make the total quantity of merchandise variable;[233] to allow for additional variables consisting of n capital services, of the total overall costs of production for the various commodities and of the overall revenue deriving from the sale of these commodities;[234] to add the equations which match, for each type of capital service, the quantity available with the quantity used and, for each commodity, the total revenues with the total cost, as well as the unit price and the unit cost of production.[235] It thus emerges that, here too, the number of equations is equal to the number of variables.[236]

1.5 Reflections on the Past and the Present of Pure Economics

Pareto recalled that in the early days of pure economics it was thought, "mistakenly, as in the case of Walras", that the simple application of mathematics to the issues of political economy would bestow on them "a rigour and a demonstrability which they were lacking" and that in this manner it would even be possible for them to obtain general acceptance.[237] The utility of mathematical analysis, however, consists only in the investigation of "certain general and qualitative problems, such as for example an exact knowledge of the conditions determining economic equilibrium".[238]

In Pareto's view, pure economics can be characterised according to its point of departure, which can be the treatment of the law of demand as axiomatic, following Cournot and Marshall, or "more profoundly, to start out from the feelings that commodities engender in human beings as objects of gratification", as Dupuit, Gossen and Jennings had started to do, followed by "Jevons, Menger, Walras, Launhardt, Marshall, Pantaleoni, Edgeworth, Fisher, Lehr and Pareto".[239] Pure economics is also distinguished for its subject matter, with Walras representing "the first to approach economic phenomena as interconnected".[240]

In the light of this, Pareto maintained that progress in the science of economics "consists in the passage from theories breaking down general equilibrium into various specific equilibria to theories addressing general equilibrium which do not break it up", acknowledging that "the use of mathematics is justified only [for the latter] while it is pointless and therefore harmful [for the former]"[241] and adding that considering "given the distribution of the commodities ... deprives the conclusions [of pure economics] of a large part of their importance".[242] In the last analysis, Pareto thought that political economy and the science of finance "are still quite unscientific"[243] and that "right now the study of concrete phenomena is more urgent" because "if observation without theory is empiricism", "theory without observation risks being pure fantasy".[244]

On a personal level, Pareto underlined, first and foremost, that "the principal difference between the Austrian school and the method adopted [by him]" consists in the fact that the former "admit one cause for the value" while, in his view, "the exchange values depend on all the circumstances associated with barter, with production and with capitalisation".[245] With this established, he pointed out that the theoretical foundation of his conception of general equilibrium had evolved from the concept of ophelimity (in the *Cours*) to that of the index of ophelimity (in the *Manuale*) to that of the index function (in the *Manuel*): he considered this evolution to represent progress in that it bespoke the replacement of "somewhat metaphysical notions [with] concepts more and more exclusively empirical".[246]

In general, for Pareto the important thing was, precisely, to underline that the fundamental distinction between economists was not in the use of mathematics but of the scientific method. Hence he felt a greater affinity with Smith than with Walras or Marshall,[247] suggesting that in the *Manual*, he was even in contrast with the latter, who sought to "completely separate economics from sociology".[248]

In regard to Walras, Pareto also reported that "if [he] had wanted to, [he] could have presented the theory of equilibrium in a manner such that it appeared to have nothing in common with Walras' version, since the same equations can be formulated differently. But this would not have been honest on [his] part".[249] Continuing to address the study of general equilibrium in conditions of competition, Pareto ascribed to Irving Fisher[250] the merit of having investigated, from the time of his 1892 work *Mathematical investigations in the theory of value and prices*, "cases involving an unspecified number of variables, because these alone can approximate reality", despite "the mathematical complications deriving from the large number of equations among which it is necessary to perform a process of complicated eliminations".

Notes

1. See (Mornati 2018b, pp. 73–99).
2. In December 1906 Pareto expressed the view that pure economics falls within the scope of the political and social sciences, Pareto to Pantaleoni, 25th December 1906, see (Pareto 1984a, p. 467).
3. Pareto to Pantaleoni, 14th December 1898, ibid., p. 253.
4. Vaud Cantonal Archives, Lausanne, Stella Association Archive, correspondence received. It is of interest to point out that at the time membership was open not only to students of engineering as now but also to those from other faculties. It is thus legitimate to conclude that Pareto considered the first public presentation of the novel theory of choice to be comprehensible and of interest also for students lacking in specialised mathematical competence, such as his students in jurisprudence who belonged to the association.
5. See (Pareto 1898, reprinted in Pareto 1987, p. 106).
6. Ibid., p. 107.
7. Ibid.
8. Pareto does not address the conditions for the maximisation of the utility function beyond the first order.
9. According to Pareto, "the order of consumption can be disregarded, because once the goods are available, they can be consumed in any desired order", Pareto to Antonio Graziadei, 16th June 1901, see (Pareto 1975a, pp. 427–428). In fact, Pareto maintained this approach and years later, in Pareto to Guido Sensini, 27th July, 1906, ibid., p. 570, he stated that his paper (Pareto 1906a) "has no significance in practice, but in theory has significant ramifications" as it indicated the mathematical conditions for the specification of the elementary ophelimities of goods, taking into account whether or not these were dependent on the order of consumption, ibid., pp. 29–30. Pareto's view was that the second category considered, that of closed cycles, which "is the most important, since we can generally assume that the individual follows the order which is most

advantageous to him, thus determining the path to be followed" (see Pareto 1911 reprinted in Pareto 1989a, p. 327). It is known that in the aftermath of the First World War, only the first part of this article was published, while the rest was lost.

10. Some time later, Pareto observed that the utility or ophelimity function Φ "always" exists only if we take into consideration solely the functions φ_a (x_a), φ_b (x_b), ... On the other hand, this function may not exist if we consider the functions $\varphi_a(x_a, x_b, \ldots)$, φ_b $(x_a, x_b, \ldots)$, ... However, he took the view that ignoring this complication "in many cases leads only to insignificant errors", see (Pareto 1902a) translated by (Sensini 1906, pp. 431–432).
11. See (Pareto 1898, reprinted in Pareto 1987, p. 108).
12. Ibid.
13. Ibid.
14. Ibid.
15. Pareto to Hermann Laurent, 7th January 1899, see (Pareto 1989b, p. 337). On this occasion Pareto, ibid., stated that at that time, (Laurent 1885–1891) was an important point of reference for him. On this episode see (Chipman 1976). It is of interest to note that shortly afterwards, Pareto to Sensini, 18th January 1905, see (Pareto 1975a, p. 533), stated that "the study of the foundations [of mathematics] is evolving and in the name, or under the pretext, of rigour, various minutiae are pursued", while underlining that "for practical purposes, all that is needed is the time-honoured knowledge, a sufficient understanding [of which] can be gained by studying" (Hoüel 1878–1881). In this same period, Pareto stated that in order to "perform original work [in mathematical economics] it is necessary to [possess a good knowledge of] differential calculus and the theory of differential equations" while in order to "simply understand works on mathematical economics some notion of differential and integral calculus suffices", Pareto to Fabio Cristiani, 14th July 1906, Banca Popolare di Sondrio-letters archive (BPS-la).
16. Pareto to Laurent, 14th January 1899, see (Pareto 1989b, p. 341).
17. A somewhat simplified version of Pareto's notation is shown.
18. Pareto to Laurent, 11th January 1899, ibid., p. 339. In fact, if we start from the consumption of r_a and continue with the consumption of r_b, we obtain the increase in utility φ_a $(r_a + dr_a, r_b) + \varphi_b(r_a + dr_a, r_b + dr_b)$, while if we start from the consumption of r_b and continue with the consumption of r_a, the generally different increase in utility φ_b $(r_a, r_b + dr_b) + \varphi_a$ $(r_a + dr_a, r_b + dr_a)$ will be obtained, Pareto to Laurent, 19th January 1899, ibid., p. 349.
19. Pareto to Laurent, 14th January 1899, ibid., p. 343.
20. Ibid.

21. Ibid.
22. Ibid., p. 345.
23. Ibid., p. 343.
24. Ibid.
25. Ibid., p. 344.
26. Ibid., p. 347.
27. Pareto to Pantaleoni, 28th December 1899, see (Pareto 1984a, p. 288).
28. Ibid., p. 292.
29. Ibid., pp. 289–290.
30. Ibid., p. 291.
31. Ibid., pp. 292–293. According to (Pareto 1900a, p. 540), these curves correspond to staple goods (i.e. those "for which our appetite tends to become a fixed quantity") and to luxury goods (i.e. those "of which one gladly chooses the greatest quantity possible ... and which one forgoes easily, rather than forgoing others").
32. Ibid.
33. See (Pareto 1900b, p. 219–220).
34. Ibid., p. 222.
35. Pareto to Pantaleoni, 23rd July 1900, see (Pareto 1984a, pp. 323–324).
36. See (Pareto 1900a, pp. 511–512).
37. See (Pareto 1900b, p. 227).
38. Ibid., p. 224.
39. Ibid., p. 223.
40. Ibid., p. 220.
41. Ibid., p. 221.
42. That is, combinations of quantities of goods.
43. See (Pareto 1900a, pp. 511–512).
44. Ibid.
45. Ibid.
46. Ibid., p. 514.
47. Ibid., p. 512.
48. Ibid.
49. Ibid., p. 548.
50. Ibid., p. 515.
51. Ibid.
52. Ibid.
53. Ibid., p. 516.
54. Ibid., pp. 515–516.
55. Ibid., p. 516.
56. Ibid.
57. Pareto shows them all as waves.
58. Ibid., p. 517.

59. Ibid.
60. Ibid., p. 518.
61. Ibid.
62. Ibid., p. 519.
63. Ibid. In relation to the curves of indifference, Pareto considered that the *maximum maximorum* "may exist or not exist, or may be placed at infinity", but even if it exists, he did not think, at first sight, "that we can imagine a mechanical attraction towards [it] in order to explain the actions of the *economic agent*", Pareto to Amoroso, 14th May 1907, see (Pareto 1975a, p. 594).
64. See (Pareto 1900a, p. 26). Pareto, ibid., pp. 543–546, also gave some specific formulations of the lines of indifference.
65. Pareto, ibid., p. 537 pointed out that from $(\partial I/\partial x)\,\mathrm{d}x + (\partial I/\partial y)\,dy = 0$ we can obtain $(\partial I/\partial x)\,(dx/dy) + (\partial I/\partial y) = 0$, from which it can be extrapolated that since $\partial I/\partial x$ e $\partial I/\partial y$ has the same positive sign, therefore $dx/dy < 0$, confirming that the curve of indifference is inclined negatively.
66. Ibid., p. 527.
67. Ibid., p. 523.
68. See (Pareto 1901, p. 236).
69. Pareto to Wilhelm Franz Meyer, 16th December 1901, see (Pareto 1975a, pp. 439–440). After this, Pareto specified that the prices "are a result and not a pre-existing element of economic reality (since) they depend on all the circumstances of the market, of production, of society, of the political context, etc.", see (Pareto 1907, reprinted in Pareto 1988, p. 56).
70. See (Pareto 1902a, p. 445). In the terms of today's microeconomics, note that the equilibrium of exchanges and production arises, for any given combination of quantities of goods produced, when the marginal rate of transformation corresponds to consumers' marginal rates of substitution: consequently, prices have disappeared.
71. See (Pareto 1902b, reprinted in Pareto 1987, pp. 129–130).
72. Ibid., p. 130.
73. Ibid.
74. Ibid., p. 131.
75. Ibid.
76. See (Pareto 1902c, p. 405).
77. Ibid., p. 407.
78. Pareto to Pantaleoni, 19th November 1898, see (Pareto 1984a, pp. 278–279).
79. Pareto to Francesco Papafava, 27th November 1888, see (Pareto 1981, p. 590).

80. As is commonly known, the original edition of this work, to which reference is principally made here, was published in Italian as (Pareto 1906b). The second edition was published in French as Pareto 1909): Pareto was far from satisfied with this translation by the socialist scholar Alfred Bonnet, Pareto to Pantaleoni, 7th and 13th September 1898, see (Pareto 1984b, pp. 86–87, 117). Recently, there have been two major critical editions: see Pareto (2006) and Pareto (2014). As regards the first uses of the volume for teaching purposes, Pareto, who continued to believe that "there is a certain interest for the University [of Lausanne] to preserve the teaching of pure economics … on a mathematical and scientific basis", Pareto to André Mercier, 24th November 1906, see (Pareto 1975a, p. 577), reported that in the course of pure economics "for this audience, [I] cannot cover the theories from the Appendix [to the Manual] but [I] cover, roughly speaking, the theories from the non-mathematical part", Pareto to Attilio Cabiati, 13th February 1908, ibid., p. 624.
81. See (Pareto 1906b, chapter III, §18).
82. Ibid.
83. Ibid., chapter III, §21.
84. Ibid., chapter III, §14.
85. Ibid., chapter III, §22.
86. Ibid., chapter III, §27.
87. Ibid., chapter III, §18.
88. Ibid., chapter III, §39.
89. Ibid., chapter III, §§46, 160, 162. An important instance of this case consists in "all transactions relating to household consumption", ibid., chapter III, §44.
90. Ibid., chapter III, §41.
91. Ibid., chapter III, §§47, 161, 162.
92. Ibid., chapter III, §42.
93. Ibid., chapter III, §51.
94. Ibid., chapter V, §9.
95. Ibid., chapter V, §10.
96. Ibid., chapter V, §10.
97. Ibid.
98. Chapter IV, §4.
99. Ibid., chapter IV, §8.
100. Ibid., chapter IV, §13.
101. Ibid., chapter IV, §8.
102. Ibid., chapter IV, §§8, 10.
103. Ibid., chapter III, §52.
104. Ibid., chapter IV, §§33, 35, 36, 39, 40.

105. Pareto, ibid., chapter III, §99, stated that the curves of indifference of the population in general can be derived from "the curves of indifference of the individuals constituting it", without, however, specifying how.
106. That is, "a decrease of x must be compensated by an increase in y and vice versa", see (Pareto 1909, appendix, §44).
107. That is, "the variable quantity *dy* which an individual is disposed to exchange along a line of indifference for a constant quantity *dx* diminishes with each increase in *x*", ibid.
108. Thus b_y, c_z, ... represent the quantities which, in today's elementary microeconomics, are termed marginal rates of substitution.
109. Ibid., appendix, §13.
110. Given that $dx = (\partial x/\partial y)dy$.
111. [110] We follow the interpretation given by Aldo Montesano, see (Pareto 2006 p. 647, note 1).
112. See (Pareto 1909, appendix, §13).
113. Ibid., appendix, §6.
114. Ibid., appendix, §19. Therefore, the integration is not possible if the total ophelimity depends on the order of consumption, of which we have no empirical knowledge (see also note 10 above); total ophelimity does not depend on the order of consumption but the ophelimity resulting from the consumption of a given quantity of a commodity depends on all the commodities available.
115. Ibid., appendix, §13.
116. Ibid., appendix, §13.

    "The relation between ophelimity and the indices of ophelimity is similar to [that] between the quantity of heat and the temperature", that is, a relation of a purely ordinal character which indicates for two differing indices of ophelimity that the corresponding quantities of ophelimity are different, but not by how much, Pareto to Giovanni De Fraja Frangipane, 18th February 1907, in Banca Popolare di Sondrio letter archive (BPS-la).
117. From an algebraic point of view both elementary ophelimities could have a negative sign too: the common positive sign is preferred because it connotes the condition adopted previously that "a combination to be preferred to another should have a greater index", see (Pareto 1909, appendix, §46).
118. Ibid., appendix, §11.
119. That is, when the two commodities are independent of each other, ibid., appendix, §47.
120. Ibid., appendix, §11.
121. See (Pareto 1906b, chapter III, §69).
122. Ibid., chapter V, §1.
123. Ibid., chapter III, §§73–74.

124. Ibid., chapter III, §60.
125. Ibid., chapter III, §96.
126. Ibid., chapter V, §21. Capital appears on a company's balance sheet "for the expenditure required to replenish it and for the cost associated with its use", ibid., §29.
127. Ibid., chapter V, §19.
128. Ibid., chapter V, §21.
129. Ibid., V, §16.
130. Ibid., chapter III, §72.
131. Ibid., chapter III, §102.
132. Ibid., chapter III, §103.
133. Ibid., chapter III, §105.
134. Ibid., chapter III, §75. An alternative formulation could be that "the locus of complete transformations is where the accounts are settled without either profit or loss", Pareto, ibid., chapter III, §176, with the implication that "if we consider the phenomenon on the average and for a sufficiently long time, in the end the greater part of the profit deriving from all the endeavour of the companies is enjoyed by the consumers", ibid., chapter V, §74. Pareto also specified, ibid., chapter V, §75, that in the transition to equilibrium, the concerns which are "quicker and more judicious will make a profit, temporarily, while the more tardy and less astute will lose out and be ruined".
135. Ibid., chapter III, §75.
136. Ibid., chapter III, §76. See also Pareto to Alfonso de Pietri Tonelli, 6th December 1912, see (Pareto 1975b, pp. 796–798).
137. At a later juncture, Pareto to Amoroso, 14th May 1907, see (Pareto 1975a, pp. 593–594), declared that the *economic agent*, having reached the point of tangency, continues his movement, as in the case of inertia in mechanics. If he continues along the route, his movement will terminate at a point situated at the same level as that from which he started, while if he continues along the curve of indifference the movement "will continue indefinitely". Experience (especially observation) identifies cases of the first type, for example, "people who suddenly earn a large profit", Pareto to Pantaleoni, 14th May 1907, see (Pareto 1984b, p. 29). Even if we accept that habit in economics corresponds to inertia in mechanics, "the difficulty, [nevertheless], is in finding out what in economics corresponds to mass in mechanics and what in economics corresponds to acceleration multiplied by mass in mechanics", so that "if this is not known to us we cannot formulate the equations of economic dynamics", Pareto to Amoroso, 14th May 1907, cit.
138. See (Pareto 1906b, chapter III, §60).
139. Which are obtained by varying the route followed.

140. Ibid., chapter III, §97.
141. Ibid., chapter III, §184.
142. Ibid., chapter III, §186.
143. Ibid., chapter III, §99.
144. Ibid., chapter III, §34.
145. Ibid., chapter III, §33.
146. See (Pareto 1909, appendix, §24).
147. Ibid., appendix, §24, underlined that "experience yields only" the ratios φ_y/φ_x, φ_z/φ_x, ...
148. Ibid., appendix, §41.
149. Ibid., appendix, §37.
150. Ibid., appendix, §38.
151. When c is a constant, ibid., appendix, §38.
152. With x_0, y_0, z_0, ... representing the initial quantities of commodities X, Y, Z, ...
153. Ibid., appendix, §41.
154. On the other hand, Pareto specified that in practice "mankind represents a very imperfect scales for weighing ophelimities", with the implication that "the equivalence of weighted ophelimities occurs only with a greater or lesser degree of approximation", see (Pareto 1906b, chapter IX, §2).
155. This definition of prices is valid only for the equilibrium values for the quantities of commodities, see (Pareto 1909, appendix, §41).
156. Making use, in particular, of matrix algebra.
157. For details see (Mornati 2018b, §§3.4, 3.5).
158. See (Pareto 1909, appendix, §§52–54).
159. Ibid., appendix, §27.
160. See (Pareto 1906b, chapter III, §100).
161. Ibid., chapter III, §101.
162. Ibid., chapter III, §116.
163. Ibid., chapter III, §117. At this point of intersection, supply and demand are equal for each commodity, ibid., chapter III, §189.
164. Ibid., chapter III, §128.
165. Ibid., chapter III, §123.
166. Ibid., chapter III, §124.
167. Ibid., chapter III, §118.
168. Ibid., chapter III, §130.
169. Ibid.
170. See (Pareto 1909, appendix, §29).
171. Ibid., appendix, §20.
172. Where μ is a parameter.
173. Ibid., appendix, §34.
174. See (Pareto 1906b, appendix, §26).

175. Ibid., appendix, §30.
176. Ibid., appendix, §26.
177. The addends containing zero in the subscript indicate initial quantities, while those whose addends do not contain zero indicate quantities at any other moment of the exchange, such as the state of equilibrium.
178. Ibid., appendix, §27.
179. Ibid.
180. Ibid., appendix, §30.
181. Ibid., appendix, §28.
182. Ibid., appendix, §63.
183. Ibid., appendix, §30.
184. Readers are reminded that Pareto, in the course of his long managerial experience, was involved at an executive level with situations of non-free competition, see (Mornati 2018a, chapter III).
185. See (Pareto 1909, appendix, §67).
186. More precisely, the equation $\varphi_{1x} = \varphi_{1y}/p_y$ is discarded.
187. Ibid., appendix, §68.
188. Pareto, ibid., appendix, §76, implicitly introducing the concept of monopolistic competition and possibly drawing on his experience in management, observed that "there are certain additional circumstances, relating to credit, customer care, etc. which can serve to differentiate commodities which are otherwise identical".
189. Ibid., appendix, §69.
190. Ibid.
191. Ibid., appendix, §76.
192. Ibid., appendix, §71.
193. Ibid.
194. Ibid., appendix, §77.
195. A is supposed to be the *numéraire*, with the implication that $p_a = 1$.
196. The only context we will deal with, for the sake of simplicity.
197. If the enterprises in free competition sold less than they produced, they would incur an operating deficit.
198. See (Pareto 1906b, appendix, §39).
199. See (Pareto 1909, appendix, §78).
200. Ibid., appendix, §79.
201. See (Pareto 1906b, appendix, §37).
202. Ibid., appendix, §37.
203. Ibid., chapter III, §106.
204. Here the selling price (corresponding to the inclination of the curve of exchanges) and the cost of production (corresponding to the inclination of the curve of complete transformations) are identical, ibid., chapter III, §194.

205. Ibid., chapter III, §107.
206. Ibid., chapter III, §151.
207. Ibid., chapter V, §7.
208. See Furlan (1908). For a short biographical sketch, see (Mornati 2006).
209. Pareto to Furlan, 11th October 1907, see (Pareto 1975a, p. 609).
210. See (Pareto 1913a) and Pareto (1917–1919, §2134).
211. See (Mornati 2018b, pp. 111–121).
212. Pareto to Sensini, 22nd March 1912, see (Pareto 1975b, p. 764).
213. See (Pareto 1911, §1).
214. Ibid., §2.
215. Ibid.
216. Ibid.
217. On the construction of these, ibid., §§3–4.
218. Ibid., §3.
219. Ibid., §8.
220. Ibid., §7.
221. Ibid., §8.
222. Ibid.
223. Ibid., §9.
224. Ibid., §10.
225. Ibid.
226. Ibid., §11.
227. Ibid., §27.
228. Ibid.
229. Ibid., §28.
230. Ibid., §31.
231. Pareto, ibid., indicated that the budget equation for individual $\theta + 1$ is not specified as it is a consequence of the remaining budget equations and of the m equations of the total (and constant) quantities of the commodities exchanged.
232. The general equilibrium of exchange with two monopolists, each of them having a monopoly of a single commodity, is likewise determined, ibid., §38, while the general equilibrium of exchange with two monopolists of the same commodity has more equations than variables, ibid., §37.
233. Ibid., §41.
234. Ibid., §42.
235. Ibid., §46.
236. Ibid. The general equilibrium of production and of exchange also includes that of capitalisation as this is interpreted as the production of new capital services, ibid., §47 and is thus conceptually equivalent to "the manufacture of ordinary items", ibid., §48.
237. See (Pareto 1913b, reprinted in Pareto 1987, p. 164).

238. See (Pareto 1902a, pp. 426–427).
239. Ibid., p. 425.
240. Ibid., pp. 426–427.
241. Pareto to Pantaleoni, 28th September 1907, see (Pareto 1984a, p. 64).
242. See (Pareto 1902a, pp. 426–427).
243. Pareto to Sensini, 5th April 1917, see (Pareto 1975b, p. 958).
244. Pareto to Pietri Tonelli, 8th August 1917, see (Pareto 1975b, p. 978).
245. Pareto to Graziadei, 29th March 1901, see (Pareto 1975a, p. 428); Pareto to Domenico Berardi, 18th August 1901, see (Pareto 1989b, p. 136).
246. Pareto to Amoroso, 11th January 1909, see (Pareto 1975a, p. 849).
247. Pareto to Felice Vinci, 16th January 1912, see (Pareto 1975b, p. 757).
248. Pareto to Vinci, 16th January 1912, see (Pareto 1975b, p. 759).
249. Pareto to Sensini, 8th August 1911, see (Pareto 1975b, p. 735).
250. See (Pareto 1911, §31, note 44).

Bibliography

Chipman, John Somerset. 1976. An episode in the early development of ordinal utility theory, Pareto's letters to Hermann Laurent. *Revue européenne des sciences sociales* XIV (37): 39–64.

Furlan, Luigi Wladimiro. 1908. Cenni su una generalizzazione del concetto di ofelimità [Notes on a generalisation of ophelimity]. *Giornale degli Economisti* XIX (XXXVII-3): 259–265.

Hoüel, Jules. 1878–1881. *Cours de calcul infinitesimal [Course in infinitesimal calculus]*. Paris: Gauthier-Villars.

Laurent, Hermann. 1885–1891. *Traité d'analyse [Treatise on mathematic analysis]*. Paris: Gauthier-Villars.

Mornati, Fiorenzo. 2006. Luigi Wladimiro Furlan. In *Dictionnaire Historique de la Suisse*, vol. 5, p. 273. Hauterive: Attinger.

———. 2018a. *An intellectual biography of Vilfredo Pareto, I, From Science to Liberty (1848–1891)*. London: Palgrave Macmillan.

———. 2018b. *An intellectual biography of Vilfredo Pareto, II, Illusions and delusions of liberty (1891–1898)*. London: Palgrave Macmillan.

Pareto, Vilfredo. 1898. *Comment se pose le problème de l'économie pure [Expounding the pure economics]*. Lausanne: self-publication.

———. 1900a. Sunto di alcuni capitoli di un nuovo trattato di economia pura-II [Summary of some chapters from a new treatise on pure economics—II]. *Giornale degli Economisti* XI (XX-6): 511–549.

———. 1900b. Sunto di alcuni capitoli di un nuovo trattato di economia pura-I [Summary of some chapters from a new treatise on pure economics—II]. *Giornale degli Economisti* XI (XX-3): 219–235.

———. 1902a. Anwendungen der Mathematik auf Nationalökonomie [Applications of mathematics to political economy]. In *Encyclopädie der mathematischen Wissenschaften mit Einschluss ihrer Anwendungen [Encyclopaedia of mathematical science and its applications]*, vol. I, section 7, pp. 1904–1112. Leipzig: Teubner.

———. 1902b. *L'économie pure [Pure economics]*. Summary of the course given at the École des Hautes Études Sociales di Paris. Lausanne: self-publication.

———. 1902c. Di un nuovo errore nell'interpretare le teorie dell'economia matematica [On a new error in the interpretation of the theory of mathematical economics]. *Giornale degli Economisti* XIII (XXV–5): 401–433.

———. 1906a. L'ofelimità dei cicli non chiusi [The ophelimity of open cycles]. *Giornale degli economisti* XVII (XXXIII–1): 15–30.

———. 1906b. *Manuale d'economia politica con una introduzione alla scienza sociale [Manual of political economy with an introduction to social science]*. Milano: Società Editrice Libraria.

———. 1907. Metafisica economica [Economic metaphysics]. *La libertà economica*. 15th–30th November.

———. 1909. *Manuel d'économie politique [Manual of political economy]*. Paris: Giard et Brière.

———. 1911. Economie mathématique [Mathematical Economics]. In *Encyclopédie des sciences mathématiques [Encyclopaedia of Mathematical Sciences]*, tome I, vol. IV, pp. 591–640. Paris: Gauthiers-Villars.

———. 1913a. Il massimo di utilità per una collettività in Sociologia [The maximum of collective ophelimity in Sociology]. *Giornale degli Economisti* XXIV (XLVI–4): 339–341.

———. 1913b. *Introduction* to Antonio Osorio. In *Théorie mathématique de l'échange [The mathematical theory of exchange]*, V–XVIII. Paris: Giard et Brière.

———. 1917–1919. *Traité de sociologie générale [Treatise on general sociology]*. Lausanne-Paris: Payot.

———. 1975a. *Epistolario 1890–1923 [Correspondence, 1890–1923]*. Complete works, tome XIX–1, ed. Giovanni Busino. Geneva: Droz.

———. 1975b. *Epistolario 1890–1923 [Correspondence, 1890–1923]*. Complete works, tome XIX–2, ed. Giovanni Busino. Geneva: Droz.

———. 1981. *Lettere 1860–1890 [Letters 1860–1890]*. Complete Works, tome XXIII, ed. G. Busino. Geneva: Droz.

———. 1984a. *Lettere a Maffeo Pantaleoni 1897–1906 [Letters to Maffeo Pantaleoni 1897–1906]*. Complete works, tome XXVIII.II, ed. Gabriele De Rosa. Geneva: Droz.

———. 1984b. *Lettere a Maffeo Pantaleoni 1907–1923 [Letters to Maffeo Pantaleoni 1907–1923]*. Complete works, tome XXVIII.III, ed. Gabriele De Rosa. Geneva: Droz.

———. 1987. *Marxisme et économie pure [Marxisme and pure economics]*. Complete works, tome IX, ed. Giovanni Busino. Geneva: Droz.

———. 1988. *Pages retrouvées [Rediscovered pages]*. Complete Works, tome XXIX, ed. Giovanni Busino. Geneva: Droz.

———. 1989a. *Statistique et économie mathématique [Statistics and Mathematical Economics]*. Complete Works, tome VIII, Geneva: Droz.

———. 1989b. *Lettres et Correspondances [Letters and correspondence]*. Complete works, tome XXX, ed. Giovanni Busino. Geneva: Droz.

———. 2006. *Manuale di economia politica [Manual of political economy]*. Edited by Aldo Montesano, Alberto Zanni and Luigino Bruni. Milano: Università Bocconi Editore.

———. 2014. *Manual of political economy. A critical and variorum edition*. Edited by Aldo Montesano, Alberto Zanni, Luigino Bruni, John S. Chipman and Michael McLure. Oxford: Oxford University Press.

Pareto, Le nuove teorie economiche. Appunti, «Giornale degli Economisti», XII (1901), XXIII-3, pp. 235–252: 236.

Sensini, Guido. 1906. Applicazioni della matematica all'economia politica del Prof. Vilfredo Pareto [Applications of mathematics to the political economy of Prof. Vilfredo Pareto]. *Giornale degli economisti* XVII (XXXIII-5): 424–453.

CHAPTER 2

Arguments in Applied Economics

In 1905, in the *Manuale di economia politica* (where the proportion of the text dedicated to applied economics is much smaller compared to the *Cours d'économie politique*, with 30% as compared to 80%), Pareto noted that since "the phenomena studied in pure economics diverge from real-world phenomena ... it would be a vain and unreasonable ambition to expect to interpret the latter by reference only to the theories of pure economics",[1] adding that, in general, "increasingly, the economic system tends to be managed in accordance with the interests of the social classes which dominate the government".[2] Such considerations helped to open the way towards the wide-ranging and innovative form of sociology which we will examine in Chap. 5; nonetheless, we consider it worthwhile to give an outline here of the practical features of the economic system which attracted the attention of Pareto the applied economist, even if they can be fully understood only from the perspective of Pareto's very particular brand of sociology. These arguments constitute a development of themes he had already explored in greater or lesser depth, including the theory of international commerce, income distribution, economic crises, the demographic issue, progressive taxation and social welfare. However, there are also a number of new areas such as the notion of the maximum of collective ophelimity in sociology, an outline of the sociological conception of savings, the statistical study of the relationship between economic and social systems, the management of public debt, the possible cooperative management of the Italian railways and strikes.

F. Mornati, *Vilfredo Pareto: An Intellectual Biography Volume III*,
Palgrave Studies in the History of Economic Thought,
https://doi.org/10.1007/978-3-030-57757-5_2

2.1 A Non-free Trade Formulation of the Theory of International Commerce

Possibly the most interesting of the topics in applied economics covered by Pareto in the *Manuale di economia politica* (and in *the Manuel d'économie politique*), for the major conceptual developments proposed, is the theory of international commerce, whose importance in the development of Pareto's economic thinking in his youth as well as during the Lausanne period we have already explored extensively.[3]

Various countries[4] are considered in the field of international commerce. If, for the sake of simplicity, we consider only two of these, then the problem in economics consists in the determination not only of the general equilibrium for each country but also of the quantities of the n commodities exchanged between them and of the exchange rates between their currencies.[5] Corresponding to these variables are the same number of equations by reference to which they can therefore be specified,[6] that is, the equilibrium balance of trade for one country (the balance of trade for the other country being identical but with the signs inverted)[7] and the prices of the n commodities exchanged.[8]

On this basis, Pareto presented Ricardo's theory of comparative costs as follows:[9] if, in a day, workers I and II are able to produce the following quantities of products A and B

	I	II
A	$\frac{6}{5}$	1
B	$\frac{4}{3}$	1

then by engaging for a period of 30 days in both categories of production, they will obtain the following quantities, which we assume are sufficient to satisfy their preferences:

	I	II	Total quantity
A	36	30	66
B	40	30	70

Following the familiar reasoning of Ricardo, it will be seen that not only is I more efficient than II in both types of production, but that he is so to a greater degree in the production of B compared to that of A, since 4/3 is greater than 6/5. Hence, since I (II) has a comparative advantage over II (I) in the production of B (A), I (II) will devote all the 60 days of work at his disposal to the production of B (A) alone, procuring the following quantities, with I exchanging B for A with II and vice versa

	I	II	Total quantity
A		60	60
B	80		80

However, since the total quantity of B has increased (80 against 70) and the total quantity of A has decreased (60 against 66), it is not possible, in general, to establish "whether, in the light of the individuals' preferences, there is compensation or not". Ricardo's theory is thus corroborated only if the quantities of A and B produced by I and II when each produces a single commodity (i.e. in conditions of free trade) are greater than the total quantities of the two commodities produced when I and II both produce them (i.e. in conditions of autarky).[10]

The following definitions also apply:[11]

in conditions of free trade, let "a" represent the quantity produced and sold domestically and "p" the price; following the introduction of an import tariff,[12] let "b" represent the quantity produced and sold domestically at a new price "p′", "c" the quantity produced domestically and sold abroad at a price "p″" and "q" the unit cost of production for "$b + c$".

In an import tariff regime, the producers will obtain a net profit if

$$\mathrm{p}'\mathrm{b} + \mathrm{p}''\mathrm{c} > (b + c)q$$

In a free-trade regime, assuming that they produce the same quantity "$b + c$" at a unit cost "q" and sell "a" domestically at a price "p" and the remainder (b + c − a) abroad at a price of "p″", the producers will not gain a net profit if

$$pa + (b + c - a)\mathrm{p}'' < (b + c)q$$

Again, in an import tariff regime, any net profit for the producers will be greater than the losses for consumers (which will always occur since, by definition, "p′" is greater than "*p*"), if

$$\mathrm{p'b} + \mathrm{p''c} - (b + c)q > (\mathrm{p'} - \mathrm{p})b$$

or alternatively if

$$pb + \mathrm{p''c} > (b + c)q$$

So,

$$pb + \mathrm{p''c} > (b + c)q \text{ and } pa + (b + c - a)\mathrm{p}'' < (b + c)q$$

imply

$$pb + \mathrm{p''c} > \mathrm{pa} + (b + c - a)\mathrm{p}''$$

or

$$(p - \mathrm{p}'')(b - a) > 0$$

which is impossible because while it is plausible that the selling price abroad p″ is lower than the domestic free-trade price "*p*", nevertheless the increase in the domestic selling price due to the import tariff will certainly reduce domestic sales of the commodity, that is, cause "*b*" to be lower than "*a*". In this manner it is impossible, in a tariff regime, for any net profit for the producers to be greater than the losses for consumers

On the other hand, customs protection, even if economically damaging, can still be useful to society by facilitating social mobility, thus promoting the emergence of new forms of social system which can preserve wealth.[13] On this basis Pareto reached the conclusion, in contradiction with one of his customary arguments,[14] that it is not possible to say in general whether "free trade or protectionism is more beneficial", a question which can only be posed, for any given country, by taking into consideration "[its] overall social and economic conditions".[15]

Within this broader theoretical perspective, Pareto emphasised that the people in favour of customs protection were those who obtained a direct benefit from it (landowners and entrepreneurs), the politicians (who could make use of the additional tax revenues to finance increases in public expenditure), the nationalists (when they imagined that "protectionism serves to defend the country from the foreigner") and those well-to-do folk who did not want "to be the only ones paying taxes" (since customs tariffs represented a cost for all the citizens, while the now-prevalent direct taxes always affected wealthy taxpayers most).[16]

Having reiterated that protectionist measures succeed because they bring (considerable) benefit to a small number of individuals, who therefore do everything they can to promote them, while the many who suffer from them suffer only negligibly (and are thus not motivated to oppose them),[17] Pareto declared that this pursuit of personal interest on the part of those benefitting from customs protection "could be for the good of the country".[18]

2.2 Further Reflections on Wealth Distribution: Criticisms of Proposed Reforms and New Implications for the Income Distribution Curve

In his *Socialist systems* (*Systèmes Socialistes*) Pareto repeated once and for all that all economic production has need of capital, labour and land, reducing to sophistry[19] any attempt objectively to identify the component due to each of these productive resources (which in turn are "in fact extremely complex and variegated"[20]) and that it is only arbitrarily that, with Marx, one can expect to assign the right to ownership of everything produced to one of these factors of production alone.[21]

Having admitted that existing ways of distributing wealth are only "the least bad" identified to date,[22] Pareto added, however, that reformers needed to formulate proposals that were "intelligible, clear and precise",[23] that were "compatible with the human character as we know it",[24] that could be achieved through "internal forces within [society], hence the need to demonstrate the existence of such forces",[25] that, once put into practice, could generate "human happiness", as subjectively understood by humanity and not by the reformers themselves.[26]

Pareto then reminded his readers that "since distribution and production are tightly interconnected", any possible modifications to them will have a reciprocal effect,[27] which suggests the need, when a modification to the method of distribution diminishes production ("and this alas is the most common case"[28]), to determine whether there are other compensating factors.[29] Lastly, Pareto observed that "distribution according to pain would certainly reduce production to a minimum" because it would discourage each person from doing the work they know best, as it would be the least remunerative since, by definition, it could be performed with a minimal effort.[30]

As for the curve of income distribution, Pareto thought it represented an "image of society", whose "outward form varies little" while the "inside is in perpetual movement", with certain individuals enriching themselves and others becoming impoverished.[31] According to Pareto, social selection operates for incomes somewhat above the minimum level for survival. This is the breeding ground of "future aristocracies" which in turn, however, are destined to crumble, possibly because their earnings will be "sufficient for them to be able to shelter" their less-able members.[32] In fact, if the eradication of these latter is prevented, together with the rise "of the most able individuals", "a revolution is imminent".[33]

2.3 Economic Crises

Economic crises always attracted Pareto's attention,[34] even if essentially only from an empirical standpoint.[35] Hence, in January 1900, he attributed the paradox of the squandering of "fabulous sums" (due to the Anglo-Boer war, strikes, municipal and state socialism) with the simultaneous increase in wealth to the fact that society was "at the end of the phase of growth and prosperity of the crisis".[36] Then, in the following May, he anticipated that the crisis just commencing in the United States would arrive in Europe "within a year at the most". Again, in December 1906, Pareto observed that the prolonged and continuing rise in all the economic indicators suggested that the economy was "in the middle of the growth phase of the crisis" so that "probably we are approaching the end of this phase".[37] In June 1907 he remarked with satisfaction that the downturn had started, just as he had recently predicted, and precisely in the nation (the United States) for which a continuing period of ascendancy had been expected,[38] thanks to the activity of its trusts.[39]

In general terms, Pareto thought that people were wrong to believe that "the period of prosperity preceding a crisis ... was going to last forever", because "crises are nothing other than the rhythmic activity of the economy, which depends on the inclinations of mankind".[40] In particular, "it is excess of confidence and rising markets which trigger excesses of pessimism and falling markets; and it is again this latter excess which replicates the first and so on indefinitely",[41] at the same time specifying that such excesses originate, in turn, with "real changes in conditions of production" together with the "subjective synchronism of economic processes".[42] Furthermore, since they emerge out of divergences between consumption and production (positive during upturns and negative during downturns), crises could only be avoided if it were possible to perfectly forecast both consumption and production,[43] which "a whole socialist regiment would probably miscalculate much more frequently than ordinary citizens" because it would be performed by clerks, while in capitalism it is performed by producers and traders who "are rewarded with riches if they guess right and with ruin if they guess wrong".[44]

On the other hand, in consolation, Pareto repeated that, firstly, at least "part of the progress achieved" during an upturn remains permanently as an asset to society;[45] secondly, that the economy could free itself of crises only "at the cost of its complete decadence and demise",[46] implying that in the end crises could be harbingers of "greater benefits than losses"[47] and that in any case, all that could be hoped for would be the attenuation of the intensity of both the downturns and the upturns in this cyclical pattern, as indeed was partially accomplished in the course of the nineteenth century.[48]

2.4 New Thoughts on the Demographic Phenomenon

Pareto also added some new thoughts on demography,[49] another of the topics that had interested him throughout his intellectual life. Having reiterated Malthus' argument concerning the obstacles acting to limit the reproductive capacity of humanity, Pareto remarked that on occasion a restraint can lead to an increase in the disproportion between population and resources, by reducing the latter to a greater extent than the former, as in the case of wars,[50] or by increasing the incentive for procreation, as with emigration or abortion, even if "there is no evidence allowing us to provide rigorous proof" of this.[51] Pareto then observed that in general, the largest possible rises in the population "benefit the rich and the political oligarchies" because they supply cheap labour to the former and bestow additional power on the latter, with the increase in the "number of subjects". Nevertheless, in practice, the rich practise birth control "in order to maintain their patrimony intact [for their children]" while the poor, who would have a motive for birth control, tend to "have a large number of children, either in order to exploit them or simply out of improvidence".[52] Furthermore, if it is believed that the revolution would be more likely to arrive "when the popular classes are oppressed by want", those in power will be in favour of birth control and the revolutionaries will oppose it,[53] while if it is thought that the arrival of the revolution would be prompted "when the masses are galvanised by affluence", the revolutionaries will favour birth control and those in power will resist it.[54] Pareto contended, however, that people's ideas about demography appeared to have no influence whatsoever on their behaviour. In fact, in France at the time, when birth control was being promoted the population increased, while when attempts were made to encourage the growth of the population it remained unchanged.[55]

2.5 Progressive Taxation

On the 30th of November 1899 Pareto[56] disparaged the previous day's successful move by the Vaud Grand Council's majority[57] (made up of radicals and socialists and vainly opposed by the liberals), with the support of the canton's government, to reform the cantonal direct inheritance tax on a more onerous progressive basis,[58] in line with the new rates being adopted for the corresponding tax levied at the municipal level.[59] Generalising from this case, Pareto claimed that the purpose of this manoeuvre was simply to increase the tax revenue to be divided up among those in the parliamentary majority, which in Italy corresponded to "the rich plunder[ing] the poor [and in Lausanne to the] poor plunder[ing] the rich".[60] Designating the plunderers as thieves and the victims as losers, Pareto stated that he "did not wish to be either a plunderer or a loser" but since this was impossible, he was seeking "somewhere to be plundered more modestly".[61]

In July 1903, Pareto pronounced that the replacement of the indirect form of taxation with the direct progressive form had originally been "a just cause" as the indirect tax amounted to a regressive direct one.[62] On the other hand, continuing increases in public expenditure had once again necessitated a substantial increase in the revenue from indirect taxes because these were "the only truly productive ones".[63] However, this increase required governments to create allies whom they could bribe by leaving them a part of the additional tax proceeds in the form of customs tariffs.[64] More generally, he thought that in the science of finance "everything remains to be done [since] it is not yet even an art", adding that the behaviour of the taxpayers, who ignored many of the tax laws, fell within the category of illogical actions "which are much more difficult to explain" than logical actions.[65]

Of the mechanisms of government, according to Pareto, we have an approximate idea of the costs, including tax takings among other things, but it is extremely difficult to gauge its output. One very loose solution to this problem is to "automatically balance the value of production against its cost".[66] Treatments of the science of finance in terms of pure economics "deviate so far from reality as sometimes to have no connection with it",[67] with the reasons for this being as follows: firstly, "the concept of needs referred to a group could have no sense except insofar as it can be considered as a single person";[68] secondly, "non-logical actions play a very important role in the phenomenon of the satisfaction of group needs".

2.6 State Social Security and Mutual Aid

Another of the topics which continued to attract Pareto's attention was social security, where he contrasted the detested variety provided by the state with his cherished mutual form, observing concrete developments in this area at both federal and cantonal level.

On the 5th of October 1899, the federal assembly (the bicameral parliament of Switzerland) gave its definitive approval for the *Federal law on insurance against sickness and accidents and on military insurance (Loi fédérale sur l'assurance contre les maladies et les accidents et sur l'assurance militaire)* which, however, was rejected by 70% of the voters in the referendum of 20th May 1900. Pareto, who had thought it better, as a foreigner, not to make any public pronouncement on the issue despite invitations to do so,[69] spoke privately at the invitation of Émile Paccaud, director of the Banque Cantonale Vaudoise (BCV) and member of the liberal faction within Vaud's radical party, who were still the dominant party at both the cantonal and the federal level.[70]

According to Pareto, on the basis of his own actuarial research and in the context of the abovementioned federal law, a provident fund directly managed by BCV for the benefit of its own employees would be able to count only on contributions made by these employees and by the bank, together with compound interest. Thus, to pay a pension for an employee who had entered the bank at 21 years of age and was now retiring at 52, the fund would have 10,647 francs available. Consequently, since the current value of a pension of 1 franc, payable each quarter to a person of 52 amounted to 12.3247 francs, this would permit payment of an annual pension of just 863 francs instead of the 1750 francs specified by the law and corresponding to half of the final salary received: "And so we either have to find new resources, or reduce the pensions".

Pareto persisted in advocating the financing of pensions through private mutual aid funds.[71] Thus, in March 1901, he offered his services to the Lausanne-based mutual aid society *Winkelried* for the purpose of examining the condition of its balance sheet, declaring that for him it was "a matter of principle to perform such work free of charge" since "when we can, we should give each other a hand".[72] The first step in determining the system of contributions needed to ensure the survival of the business was to investigate the historic workings of the mechanism. Hence, Pareto[73] calculated the average number of sick days per member in the period between 1889 and 1900 and found widely varying averages, as is normal for a society with a small number of members. Nonetheless, an attempt can at least be made to determine "the trend of these averages" for the membership as a whole, calculating both the average over the entire period 1889–1900 (where it amounted to 5.048 days) and that for the two subperiods 1889–1894 (5.12 days) and 1895–1900 (4.484 days). Since these averages did not appear to him to differ very much, Pareto concluded that if the higher figure were adopted as an average, "the society should find itself in a position of stability". Then, having demonstrated that over the period 1897–1900, collective medical and associated expenditure for the membership averaged between 1.2 and 2.2 francs, Pareto advised that it would be "prudent to allow for" an average daily medical expenditure per day of sickness corresponding to 1.9 francs, which, taken together with the 1.0 franc subsidy, yielded an overall average cost per day of sickness of 2.9 francs which, multiplied in turn by the average of 5.12 days of sickness per member, would lead to an annual expenditure per member of 14.85 francs compared to the current annual contribution of just 12 francs.

Hence, it would be necessary to raise the monthly contribution per member from 1 franc to 1.2 francs.

In July 1901, Pareto informed the President of the Bienne mutual aid society *La Fraternité* that in order to preserve its "existence ... it is indispensable to introduce variable contributions according to the age of each member", for example, by charging high entrance fees for elderly aspiring members, with monthly contributions varying with changes in members' ages.[74]

2.7 The Maximum Ophelimity for a Collectivity in Sociology

In his *Manuel d'économie politique*, Pareto, having repeated his by now well-established definition for the maximum of collective ophelimity,[75] remarked that in a context of positive general expenses, maximum ophelimity can only be achieved by a collectivist society, as this represents the only form of social organisation capable of covering general expenses through taxation and later selling the individual units of the commodities at their marginal cost.[76] The maximum of ophelimity can be achieved under free competition only if general expenses are nil and prices are variable.[77] On the other hand, Pareto grasped that all consortia (whether of capital or of labour) seek to impose production coefficients diverging from those maximising ophelimity for society, even if "most people, [out of] humanitarian zeal", object only to the ones made up "of the hated capitalists and even more hated speculators".[78] Pareto further added that in general, whatever form of economic organisation is adopted by society, people either produce goods themselves or they appropriate those of other people, and that the second alternative, while certainly militating against the achievement of the economic maximum, does not necessarily prevent the achievement of the maximum "of social utility, since the struggle to take possession of other people's possessions can facilitate the selection" of the social *élite*.[79]

A number of differing sociological maxima may exist: thus, the maximum quantity of the population conforming to the maximum of collective utility is greater, for reasons of "military and political power", than the maximum quantity of the population conforming to the maximum of collective utility, when determined on the basis not only of political and military advantages but also of the economic costs for each social class of each individual increase in population.[80]

2.8 A Note on the Sociological Conception of Savings

In April 1909 Pareto declared that Irving Fisher was mistaken in his claim that "savings are constituted in view of the fruits of what is saved". In truth, "the accumulation of savings is largely instinctive, ... It is a [self-prompted] non-logical act and people either have or don't have a taste for [it]".[81] Thus, "the consideration of the fruits becomes preponderant, together with the idea of a safety measure, [only] in determining how existing savings are utilised", implying that "for the purpose of studying economic equilibrium it is necessary to take the quantity of savings as given".[82] Savings represent a case "where the logical actions contemplated by economics must give way to the non-logical actions of sociology".[83]

2.9 A Statistical Analysis of the Relations Between Economic and Social Phenomena

Again in the context of his increasing conviction regarding the need, where possible, for a combined treatment of economic and social phenomena, which will be explored more fully in Chap. 5, Pareto also determined to embark on a statistical treatment of the relations between them.[84]

In the autumn of 1913, from his analysis of the pattern of French foreign trade between 1799 and 1910, Pareto concluded that it displayed accidental variations ("very slightly interrupting the curve, which immediately returns to its former trajectory") over short periods (with ascending and descending sections) or over longer periods (to be investigated "by seeking, by means of interpolation, to determine around which line the resulting curve oscillates").[85] Given that similar oscillations also occurred in the cases of England, Belgium and Italy, Pareto applied the method of interpolation on the historical data, finding that in these countries between 1854 and 1872 there was "a phase of rapid growth", between 1873 and 1897 there was "a period of slow growth (in Italy decline)" and from 1898 to 1913 "another phase of rapid growth which is possibly drawing to a close, since already the premonitory signs are apparent".[86] Having said this, Pareto remarked that "periods of rapid economic growth are much less agitated, from a political and social point of view, than those of economic depression", once again mentioning, for example, that the beginning of the period 1854–1872 coincided, in France, with the encouraging stability which characterised the inception of the Second Empire, while the period 1873–1897 "was likewise a time of serious social disturbances".[87]

Pareto underlined that in general, "a rapid rise in economic prosperity", or better still a victorious war, increases the self-esteem of the populations concerned, sometimes manifesting itself in political developments such as the rise of nationalism or imperialism,[88] as in the United States and in various European countries at the time.

Moreover, "rising economic prosperity" allows governments to bring forward expenditure which can be paid for later, in particular for the purpose of suborning the "chiefs" and, even more importantly, the "soldiers" of the opposition parties.[89] However, if, at the time when these payments fall due (i.e. the repayment of the debts incurred to finance the expenditure), there were a depression, this repayment would be complicated by the unexpected slowdown in the growth both of tax revenues and of savings available to finance public debt.[90]

On the other hand, if the growth in gold production continued, then prices would likewise continue to rise. This, by reducing the real value of debts, would allow governments to increase taxes without too many protests.[91] Actually inflation would increase nominal salaries, thereby weakening the opposition of salaried workers, which would be further reduced by the fact that inflation acted to encourage "the turnover of the elites" which, apart from prompting an increase in production, "will then deprive the salaried employees of the bosses who might mobilise them to revolt".[92]

2.10 The Management of Public Debt

A new topic in applied economics which Pareto briefly turned his attention to, on the invitation of the government of Vaud canton, was the management of public debt.

Pareto's view was that the problem of where to place public debt issues (whether domestically or abroad) could be examined either by considering the issuing state as a private individual taking out a loan or from a collective viewpoint.[93]

In the first case, we consider the price at which a loan can be taken out in Lausanne in Swiss francs, whose servicing (interest and depreciation) is paid in Lausanne in Swiss francs and the price at which a loan can be taken out in Paris in French francs, whose servicing is payable in Paris in French francs. Since Pareto considered that "it appears very unlikely at the moment" that fiat money would be introduced in Switzerland (with the resulting devaluation of the Swiss franc against the French franc), the exchange rate between the Swiss franc and the French franc would very probably remain at 1.006 Swiss francs to 1 French franc so that if the loan were acquired in Lausanne for an x-price of 3.5 Swiss francs and the same loan were bought in Paris at a y-price of 3.5 French francs, the government of Vaud would collect, at Paris, 3.52 Swiss francs. In Pareto's opinion, even after making allowance for the additional bank fees which would be incurred for the arrangement of the loan in Paris, it "is probable" that for the private individual it would be advantageous to place the loan in Paris.

From a collective viewpoint, Pareto noted that if the Vaud canton contracted a loan, Swiss foreign debt would increase (negatively affecting the exchange rate of the Swiss franc) by the amount of the loan, no matter what form it took, that is, whether it was issued in Switzerland or abroad. Correspondingly, if the loan were issued in Switzerland, in order to subscribe, the subscribers would draw on their bank or savings deposits or even directly their wallets, with the result that the credit institutes and also the subscribers themselves would reduce their lending "to commerce, to industry, to agriculture" by the same amount. Since the economy's need for credit does not diminish, these various sectors, "doing what they can", would turn "more or less indirectly abroad" bringing a consequent rise in Swiss foreign debt. In the wake of the Vaud loan, the Swiss foreign debt position would deteriorate, even if the subscribers raised the money by selling foreign stock, which indeed would amount to reducing Swiss outstanding credit abroad. Given that the issue of a loan on the part of Vaud would have the inevitable effect of making the Swiss foreign debt position worse, the choice between issuing it in Lausanne or directly abroad would depend on the difference in the rate available to private individuals and that at which the Vaud canton was able to borrow abroad. Pareto deemed the latter to be lower so that, in his view, at that moment placing the loan abroad "is [the] best option" for the canton.

2.11 The Possible Cooperative Management of the Italian Railways

In June 1908, since the state's management of the Italian railways (which had begun in 1905) "is a real disaster" and a return to private management "would be very difficult to achieve now",[94] Pareto considered that management by a cooperative of railway workers "might be the only viable solution",[95] while experience suggested that a new public–private partnership would be "the worst of all the options that could be imagined for the Italian railways".[96]

On the other hand, he thought that any such cooperative needed to avoid interrupting the flow of goods and passengers with strikes and should also undertake to take over the activity "on the same terms" as the private companies which the state had replaced, thus demonstrating its ability to manage the operations at least as well as the former private concessionaires.[97]

In more detail, in Pareto's view the state control of the railways in Italy which had just commenced had not succeeded in improving on the previous system of private management because "the managers lack the great incentive of a financial return", whereas a cooperative formula could be organised in such a manner as to procure "a profit derived from assuring a good service to the public".[98] Hence he proposed, as an incentive for efficiency, that the cooperative should be provided with external capital made up of bonds and preference shares, repayable at a fixed rate through preemptive deductions from gross profits and also with internal capital distributed among the railwaymen and remunerated from the remaining profit. Pareto pointed out that "this form of organisation is common among many English companies and performs well".[99] Thus the railwaymen, since their remuneration would be dependent on the profits of the railways (increasing with increases in revenue and with reductions in expenditure), "would derive the required improvement in their condition from progress made in the transportation industry".[100] Pareto also thought that the retention of the contractual terms whereby the state was able to claim 37.5% of the gross revenues continued to be "an anti-economic mechanism .... [given] that modern industry is characterised by low margins on large production volumes". As an alternative, he proposed that the cooperative should "pay a fixed annual amount, which would be unchanged for at least ten years".[101] Having said this, Pareto considered the fact that the Swiss railways, managed on a federal level since 1903, had showed relatively unsatisfying results notwithstanding the good performance of the preceding private management and also despite the quality of the public administration in general, as well as "the effective guardianship" of the electors, confirmed that the state was not suited to the task of railway management,[102] as it entailed "on one hand an increase in bureaucratic expenses and on the other a reduction in the comforts available to the public".[103]

In May 1920, Pareto noted that state control of the railways in Italy and in France had, contrary to expectations, brought a reduction in the productivity of the railway workers (meaning a higher number needed to perform the same quantity of work), an increase in the number of strikes and the exchange of an annual profit for a loss.[104] To remedy this, Pareto still believed that the best solution was to entrust the state-controlled railways to a cooperative of railway workers as "it has the enormous advantage of bringing the selfish interests of the railwaymen into play since their earnings would vary in accordance with improvements or deterioration in the service".[105] It was difficult, "but one cannot say impossible" to organise the provision of capital on the part of the state while putting in place "safeguards against the squandering of the capital".[106] On the other hand, nationalisation, an intermediate solution between state control and cooperative (or union) management, which consisted in consigning the management of the railways to "a board of directors with representatives of all the various categories who have an interest in the railways",[107] "would increase the damage and so would have to give way, sooner rather than later, to a different system for managing this type of enterprise".[108]

2.12 Strikes

A topic which, in the period around the turn of the twentieth century, often attracted Pareto's attention, initially with a certain benevolence but later with increasing disapproval, was that of strikes, which were becoming more and more frequent at the time.

In August 1899 Pareto still held the view that in general the right of workers, including those in the public sector, to join forces "in order to sell their labour at a price and on conditions which they deem appropriate" and to suspend work (while abiding by the provisions of their employment contract) if their demands were not met, was not only "fair and equitable, but [also] the only known means of determining the price of labour in such a way as to obtain the maximum well-being for society".[109]

This consideration was formulated in protest against the Pelloux government's recent legislative decree banning strikes by employees of public services, with no compensation being offered whatsoever.[110]

However, already by July 1901, having criticised the tolerance displayed by the authorities towards the violence of the strikers because "it is not compatible with civilised living"[111] and also disagreeing with the option of a ban on strikes due to its proven ineffectiveness,[112] Pareto advocated the solution of the repression of the violence by the strikers, while safeguarding the right to strike as being one of the conditions for freedom in "wage negotiations".[113] Consequently, he lauded the position recently adopted by the Home Secretary Giovanni Giolitti in the area around Ferrara.[114]

In January 1902, Pareto, in the context of a strike threatened by the Italian railway workers, commented that in modern society it is normal to attempt to reconcile "the regular functioning of certain essential services indispensable to society" while at the same time avoiding excessive impingements on the rights of those working in these services.[115] The measures suggested for the solution of this problem included coercion (with the militarisation of the Italian railway personnel), prevention (such as the ban on unionisation which was being proposed for the French railways workers) and another (which was being mooted in England) which involved negotiating only with those trade unions which, in exchange for a recompense, agreed "to ensure the normal operation of certain services" regardless.[116]

Notes

1. See (Pareto 1906a, chapter III, §228). In January 1920, Pareto expressed his view that "economics as it is taught often has nothing of the character of a science, but is rather a simple kind of literature which serves to procure university tenures, academic honours and other privileges", see (Pareto 1920a, reprinted in Pareto 1920b, p. 289).
2. See (Pareto 1906a, chapter IX, §27).
3. See (Mornati 2018a, §6.6) and (Mornati 2018b, chapter VI).
4. See (Pareto 1906a, chapter VI, §65).
5. Ibid., chapter VI, §66.
6. Ibid., chapter VI, §67.
7. Ibid.
8. Ibid., chapter VI, §§68–69.
9. Ibid., chapter IX, §45.

10. Ibid., chapter IX, §46. Pareto explored this question with various arithmetical examples too, ibid., §§46–52.
11. Ibid., chapter IX, §54, note 1.
12. Which will also lead to a reduction in the price of the commodity in the exporting country, ibid., chapter IX, §55.
13. Ibid., chapter IX, §§55–59, 72.
14. See (Mornati 2018a, pp. 153–160) and (Mornati 2018b, pp. 123–139).
15. See (Pareto 1906a, chapter IX, §60). In this new perspective, already some time earlier Pareto had approved of the renewal of customs protection granted to German agricultural producers, deeming it "the shield of the state", see (Pareto 1902a), and he later affirmed, in January 1914, that it was also in order to determine "the economic benefit or harm" generally caused by customs protection that he wrote his *Treatise of sociology*, Pareto to Luigi Einaudi, 18th January 1914, BPS-la (Banca Popolare di Sondrio-letters archive). Thus, in essence, "free trade is now a very secondary question compared to others hanging over society, especially as regards the threatened despotism of the proletariat, which could sweep European society towards unexplored scenarios", Pareto to Edoardo Giretti, 4th May 1910, see (Pareto 2001, p. 183).
16. See (Pareto 1906a, chapter IX, §63).
17. Ibid., chapter IX, §66.
18. Ibid., chapter IX, §67.
19. See (Pareto 1901a, p. 333).
20. Ibid.
21. Ibid., pp. 335–337.
22. See (Pareto 1902b, p. 156).
23. Ibid., p. 157. But this "rarely happens", ibid.
24. Ibid., p. 158. Thus, for example, Pareto thought that workers "were right" to prefer a given fixed salary, secured (for example) by means of a successful strike, to the same salary of which, however, part was received as a share in profits, which by definition are unreliable, Pareto to Papafava, 28th May 1899, see (Pareto 1989, p. 352).
25. See (Pareto 1902b, p. 158). Pareto pointed out that contrary to the prejudices "of our good humanitarians", who thought that salaries could only increase to the detriment of profits and vice versa, in the cyclical upturn which was reaching its culmination in 1907, there had been an increase in interest rates and an "even bigger increase in salaries", suggesting that at the end of the downturn which was just about to start there would be reductions both in interest rates and salaries, see (Pareto 1907a).
26. See (Pareto 1902b, p. 158).
27. Ibid., p. 158.

28. Ibid., p. 159.
29. Ibid.
30. See (Pareto 1902b, p. 167).
31. See (Pareto 1906a, chapter VII, §18).
32. Ibid., chapter VII, §19.
33. Ibid., chapter VII, §21.
34. On Pareto's earlier considerations regarding this topic, see (Mornati 2018b, §8.4).
35. Again at the end of December 1898, Pareto had reiterated, without however providing any analytical follow-up for his conviction, that "the dynamics [of the economy] provide the only rational explanation for economic crises", see (Pareto 1898, reprinted in Pareto 1987a, p. 108).
36. Pareto to Pantaleoni, 31st January 1900, see (Pareto 1984, p. 302).
37. See (Pareto 1906b).
38. See (Pareto 1907b). In any case, Pareto thought that considering the large quantities of gold arriving on the market, the new crisis would be different from previous ones "in that the fall in commodity prices will be less marked", ibid., p. 175.
39. Pareto agreed with the theory of the French journalist Paul de Rousiers arguing that the concentration of companies in an industry was a necessary but not sufficient condition for its transition to monopoly, which can occur only in the presence of other factors, among which the backing of the state is indispensable, see (Pareto 1899a, reprinted in Pareto 1966, pp. 152–153).
40. Pareto to Pantaleoni, 28th May 1900, see (Pareto 1984, p. 312). Crises can be conceived as "undulations [which] can occur around a rising line or a falling line". He added that he feared the line was falling because, for him, the legislative hurdles continually being placed on production seemed to outweigh the technical progress being achieved in industry and "the exploitation of new lands in America, in Africa, in Asia", see (Pareto 1902c).
41. See (Pareto 1904).
42. See (Pareto 1906a, chapter IX, §78).
43. Ibid., chapter IX, §76.
44. Ibid., chapter IX, §77.
45. See (Pareto 1904).
46. Ibid.
47. See (Pareto 1906a, chapter IX, §81).
48. See (Pareto 1904).
49. On Pareto's earlier ideas regarding demography, see (Mornati 2018a, §6.5) and Mornati (2018b, §8.1).
50. See (Pareto 1906a, chapter VII, §76).

51. Ibid., chapter VII, §80.
52. Ibid., chapter VII, §82.
53. Ibid., chapter VII, §84.
54. Ibid., chapter VII, §85.
55. Ibid., chapter VII, §88.
56. Pareto to Pantaleoni, 30th November 1899, see (Pareto 1984, p. 282).
57. The Great Council (Grand Conseil), *Gazette de Lausanne*, pp. 3–4, 30th November 1899.
58. Taxable income was divided into three categories (from 1 franc to 25,000, from 25,001 to 100,000 and over 100,000) with rates rising from 1.6% to 2.4% to 3.2%, ibid.
59. Ibid.
60. Pareto to Pantaleoni, 30th November 1899, see (Pareto 1984, p. 282).
61. Ibid. On Pareto's tax position of the time, see (Mornati 2018b, chapter I, note 57).
62. See (Pareto 1903).
63. Ibid.
64. Ibid.
65. Pareto to Sensini, 9th April 1905, see (Pareto 1975, p. 543).
66. See (Pareto 1917–1919, §2269).
67. Ibid., §2271.
68. Ibid.
69. Pareto to Jean Fornallaz, 26th October 1899, BPS-la.
70. Pareto to Emile Paccaud, 8th November 1899, BPS-la.
71. On Pareto's benevolent attitude to the self-help movement in Florence and in Lausanne, see (Mornati 2018b, §2.11).
72. Pareto to the president of the Winkelried mutual aid society, 9th March 1901, BPS-la.
73. Report on the Winkelried mutual aid society, 9th March 1901, BPS-la.
74. Pareto to Albert Berthelet, 4th July 1901, BPS-la.
75. See (Pareto 1909, appendix, §§89, 128).
76. Ibid., appendix, §118.
77. Ibid., appendix, §92.
78. Ibid., chapter IX, §9.
79. Ibid., chapter IX, §19.
80. See (Pareto 1917–1919), § 2134.
81. Pareto to Amoroso, 6th April 1909, see (Pareto 1975, pp. 656–657).
82. Ibid., p. 657.
83. Ibid.
84. However, again in September 1912 Pareto declared that "unfortunately [mathematical statistics] now serve only to throw dust in the eyes" since "it is ridiculous to believe that calculation can make up for imperfect data.

First the data must be valid, then comes knowing how to use it and that is where mathematics can be of assistance", Pareto to Bodio, 19th September 1912, see (Pareto 2001, p. 200).

85. See (Pareto 1913, reprinted in Pareto 1920b, pp. 7–9).
86. See (Pareto 1913, reprinted in Pareto 1920b, p. 16).
87. Ibid.
88. See (Pareto 1913, reprinted in Pareto 1920b, pp. 17–18).
89. See (Pareto 1913, reprinted in Pareto 1920b, p. 19).
90. See (Pareto 1913, reprinted in Pareto 1920b, p. 23).
91. See (Pareto 1913, reprinted in Pareto 1920b, p. 27).
92. Ibid.
93. Pareto to Ferdinand Virieux (1855–1912, at that time a member of the cantonal government), 14th June 1899, BPS-la.
94. Specifically, because "there would be no parliament to approve it and possibly no entrepreneurs to take it on", see (Pareto 1911a, reprinted in Pareto 1987b, p. 497).
95. See (Pareto 1908a, reprinted in Pareto 1988, p. 62).
96. See (Pareto 1911a, reprinted in Pareto 1987b, p. 497).
97. See (Pareto 1908a, reprinted in Pareto 1988, p. 61).
98. See (Pareto 1908b, reprinted in Pareto 1988, p. 57). See too (Pareto 1910a, reprinted in Pareto 1987b, p. 484) and (Pareto 1910b, reprinted in Pareto 1987b, p. 489).
99. See (Pareto 1911a, reprinted in Pareto 1987b, p. 497).
100. See (Pareto 1911b, reprinted in Pareto 1987b, pp. 499–500).
101. See (Pareto 1908a, reprinted in Pareto 1988, p. 62). After the war Pareto confirmed his favourable view of the management of the railways by the railwaymen, see (Pareto 1920c, reprinted in Pareto 1987b, pp. 633–635).
102. See (Pareto 1908b, reprinted in Pareto 1988, p. 57).
103. See (Pareto 1908b, reprinted in Pareto 1988, p. 58). On the similar criticisms by the youthful Pareto regarding state interference in the management of the railway service, see (Mornati 2018a, capitolo VI, § 1).
104. See (Pareto 1920c, reprinted in Pareto 1987b, pp. 633–634).
105. See (Pareto 1920c, reprinted in Pareto 1987b, p. 634).
106. See (Pareto 1920c, reprinted in Pareto 1987b, p. 635).
107. Ibid.
108. Ibid.
109. See (Pareto 1899b).
110. See (Pareto 1899c).
111. See (Pareto 1901b, reprinted in Pareto 1987b, pp. 356–357).
112. Ibid.

113. The other two conditions being the right not to strike and the right of entrepreneurs not to accept salary demands, see (Pareto 1901b, reprinted in Pareto 1987b, p. 362).
114. See (Pareto 1901b, reprinted in Pareto 1987b, pp. 363–364).
115. See (Pareto 1902d).
116. Ibid.

Bibliography

Mornati, Fiorenzo. 2018a. *An intellectual biography of Vilfredo Pareto, I, From Science to Liberty (1848–1891)*. London: Palgrave Macmillan.

———. 2018b. *An intellectual biography of Vilfredo Pareto, II, Illusions and delusions of liberty (1891–1898)*. London: Palgrave Macmillan.

Pareto, Vilfredo. 1898. *Comment se pose le problème de l'économie pure [Expounding the pure economics]*. Lausanne: self-publication.

———. 1899a. [review of de Rousiers Paul. 1898. *Les industries monopolisées (Trusts) aux Etats-Unis (Monopoly industries (Trusts) in the United States)*. Paris: Colin]. *Zeitschrift für Sozialwissenschaft*, pp. 469–470.

———. 1899b. Le droit de grève du personnel des services publiques [The right to strike of public service employees]. *Journal des économistes*, August, pp. 170–178.

———. 1899c. Le socialisme d'Etat en Italie [State socialism in Italy]. *Gazette de Lausanne*, January 29.

———. 1901a. *Les systèmes socialistes [Socialist Systems]*, tome I. Paris: Giard et Brière.

———. 1901b. Gli scioperi e il ministro Giolitti [Strikes and Minister Giolitti]. *Il Secolo*, July 19–20.

———. 1902a. Le tarif douanier allemande [The German customs tariff]. *Gazette de Lausanne*, December 4.

———. 1902b. *Les systèmes socialistes [Socialist Systems]*, tome II. Paris: Giard et Brière.

———. 1902c. La crise économique actuelle [The current economic crisis]. *Journal de Genève*, November 26.

———. 1902d. Le problème du personnel des chemins de fer en Italie [The problem of the railway personnel in Italy]. *Journal de Genève*, January 28.

———. 1903. Protection et impôt [Protection and tax]. *Journal de Genève*, July 14.

———. 1904. Le mouvement économique et la guerre [Economic changes and war]. *Journal de Genève*, November 26.

———. 1906a. *Manuale d'economia politica con una introduzione alla scienza sociale [Manual of political economy with an introduction to social science]*. Milan: Società Editrice Libraria.

———. 1906b. La crise économique [The economic crisis]. *Gazette de Lausanne*, December 3.
———. 1907a. Crise future [Future crisis]. *Gazette de Lausanne*, November 13.
———. 1907b. Période descendante [The downturn]. *Gazette de Lausanne*, June 5.
———. 1908a. Cooperativa ferroviaria? [A cooperative railway?]. *La Ragione*, June 15.
———. 1908b. L'esercizio ferroviario. Parallelo con la Svizzera. Errori fondamentali [The management of the railways. The parallel with Switzerland. Fundamental errors]. *Il corriere mercantile*, July 12–13.
———. 1909. *Manuel d'économie politique [Manual of political economy]*. Paris: Giard et Brière.
———. 1910a. Le ferrovie. Osservazioni ai professori Einaudi e Pantaleoni [The railways. Comments addressed to Professors Einaudi and Pantaleoni]. *La Ragione*, August 3.
———. 1910b. Ferrovie e industria privata [The railways and private industry]. *La Ragione*, August 31.
———. 1911a. Il problema ferroviario [The problem of the railways]. *Il Giornale d'Italia*, January 27.
———. 1911b. *Introduzione* a Vincenzo Mercadante, *Le ferrovie ai ferrovieri [Introduction* to Vincenzo Mercadante, *The railways for the railwaymen*]. Milan: Koschitz, pp. 3–5.
———. 1913. Alcune relazioni tra lo stato sociale e le variazioni della prosperità economica [On certain relations between the social state and variations in economic prosperity]. *Rivista italiana di sociologia*, September-December, pp. 501–548.
———. 1917–1919. *Traité de sociologie générale [Treatise on general sociology]*. Lausanne-Paris: Payot.
———. 1920a. Stato economico presente [The current economic situation]. *La vita italiana*, January 15, 1920, pp. 1–16.
———. 1920b. *Fatti e teorie [Facts and theories]*. Florence: Vallecchi.
———. 1920c. Statizzazione e nazionalizzazione [State control and nationalisation]. Il *Resto del Carlino*, May 25.
———. 1966. *Mythes et Idéologies [Myths and ideologies]*. Complete Works, tome VI, ed. Giovanni Busino. Geneva: Droz.
———. 1975. *Epistolario 1890–1923 [Correspondence, 1890–1923]*. Complete Works, tome XIX-1, ed. Giovanni Busino. Geneva: Droz.
———. 1984. *Lettere a Maffeo Pantaleoni 1897–1906 [Letters to Maffeo Pantaleoni 1897–1906]*. Complete Works, tome XXVIII.II, ed. Gabriele De Rosa. Geneva: Droz.
———. 1987a. *Marxisme et économie pure [Marxisme and pure economics]*. Complete Works, tome IX, ed. Giovanni Busino. Geneva: Droz.

———. 1987b. *Écrits politiques. Reazione, Libertà, Fascismo, 1896–1923 [Political writings. Reaction, Liberty, Fascism, 1896–1923].* Complete Works, tome XVIII, ed. Giovanni Busino. Geneva: Droz.
———. 1988. *Pages retrouvées [Rediscovered pages].* Complete Works, tome XXIX, ed. Giovanni Busino. Geneva: Droz.
———. 1989. *Lettres et Correspondances [Letters and correspondence].* Complete Works, tome XXX, ed. Giovanni Busino. Geneva: Droz.
———. 2001. *Nouvelles Lettres 1870–1923 [New Letters 1870–1923].* Complete Works, tome XXXI, ed. Fiorenzo Mornati. Geneva: Droz.

CHAPTER 3

The Definitive Abandonment of Liberal Political Activism

In the last volume we sought to document the thesis of Pareto's gradual distancing from his 30-year commitment to liberal activism around the turn of the century, essentially out of his disillusionment at the results which had been achieved or which were achievable.[1]

In this chapter, we intend to trace this fairly rapid and in any case definitive process, which opened the way for a more dispassionate analysis of the political, and more generally the social, scene. To recapitulate briefly, at the end of 1906,[2] Pareto acknowledged, in a rare reference to his own personal feelings, that "a decade or so [ago], [I] started to work on applied economics … and in order to work on applied economics [I] needed to have a party and that party was the liberals". He immediately added, however, that "through my studies [I] learnt something of which the economists continue to display ignorance, which is that there is a science of economics and a scientific sociology and these, like all sciences, do not and cannot have a political allegiance and serve not to issue precepts, but simply to seek to identify the patterns behind experience".[3]

3.1 Personal History

For a large part of 1899, Pareto continued to express highly negative views on the current Italian government, led from 29th June 1898 to 24th June 1900 by General Luigi Pelloux, which was to be the last government of the reactionary period which had begun with Crispi's resignation in March

F. Mornati, *Vilfredo Pareto: An Intellectual Biography Volume III*,
Palgrave Studies in the History of Economic Thought,
https://doi.org/10.1007/978-3-030-57757-5_3

1896.[4] This government was adjudged "really dire" by Pareto, to the point that "[he] would vote for any opposition party … because, by undermining the power of the government, there is hope that [it] might become less overbearing".[5] He added[6] that no political factor (such as international prestige[7]) or economic factor (such as increasing prosperity) could compensate for the current erosion of freedoms together with the heavy burden of taxation.

In September of the same year, Pareto averred that since "in politics agreement comes not from abstract principles but from the concrete steps which seem the most pressing at a given moment", it is therefore not surprising if in Italy the true liberals, that is, the few deputies from that political grouping who voted in favour of statutory liberties and against customs protection,[8] and the socialists, being both equally opposed to Pelloux's interference with individual liberties, should ally themselves against him, despite their strongly contrasting ideas on other questions which were in fact not pertinent at that particular moment.[9] This was Pareto's final display of activism, even if it was only on a private level.

In December 1899, he acknowledged that he was "the first to say that there is nothing in the world so vain as what [I] have written in criticism of the Italian government" and that he had therefore decided to "write as little as possible".[10] He then commenced an attempt to justify this decision, as well as possibly to dignify it in intellectual terms.[11] Hence, in the summer of 1900, he claimed that the reason he no longer wanted "to play the least part in active politics"[12] was that since "science is rationality and politics, faith",[13] he was convinced that "it is not possible to be active in politics and to adhere rigorously to scientific principles", adding that "the methods required for influencing people are diametrically opposed to those for discovering the truth" because "nothing can be done to influence men without a faith, whereas for science scepticism is indispensable".[14]

Pareto also considered[15] that "nowadays we have nothing but the socialist religion or the traditional ones" and to prefer one to the other "is only a matter of taste". Consequently, he had decided to limit himself to "watching and listening", firstly because he wanted "neither the one nor the other" and secondly because "regrettably everything done against one of them only serves the other". In October 1903, Pareto, after having continued notwithstanding, for a time, to be "in favour of the persecuted against the persecutors, whomsoever they may be",[16] finally reiterated his conviction that the choice was now between socialism and the reactionaries, a choice "which is mainly a question of sentiment" and concerning which he therefore had "nothing to say".[17]

Strictly speaking, in any case, the fact of distancing himself from liberalism did not leave Pareto completely indifferent to ideological questions. Thus, in 1905, after saying that he laughed at the democratic mantras in which he, too, had believed but which now seemed like "silly prejudices",[18] he underlined that even if "now he no longer [had] a party", he had "friends in all the parties",[19] insisting, however, that he had not become "a bourgeois conservative"[20] but rather an anti-humanitarian,[21] in the sense that he believed "that force alone controls the world", implying that anyone who, like the contemporary bourgeoisie, expected to be able to retain power without force and feared "to shed blood, must be eradicated and deservedly so".[22]

His disengagement from active politics was counterbalanced, in his view, by an improved capacity to understand the social system, especially as he was convinced "more and more strongly that politics is an activity for the ignorant"[23] because "in reality we know nothing about social problems and makeup, so we have no basis for our [politicising]".[24] Thus, back in January 1900, Pareto had claimed to observe "events as [I] would observe the movements of a population of ants"[25] and in August 1901, having reiterated that "increasingly with each passing day [I] shrink further from wishing to play any part in the human comedy", he observed that "in the past when [I] attempted, vainly as it turned out, to take part, all sorts of prejudices and passions clouded [my] sight, whereas nowadays [I] enjoy a vision of the truth less obscured by clouds".[26] So, in June 1905, Pareto confirmed that he sought to identify "only the scientific relations between phenomena [for] the pleasure of knowledge and to be able to write about [them] to [my] own satisfaction, but without the least intention of persuading anyone".[27] In March 1907, Pareto stated that he considered himself to be "dispassionate, precisely because [I] judge things independently of the [humanitarian ethos]" which, at that time, seemed to him to "dictate everyone's thought patterns".[28]

3.2 Final Empirical Observations on Liberalism

For a certain period, however, Pareto continued, ever more critically, to follow the liberal movement. In February 1901,[29] having remarked that "everyone is a liberal, in principle", he called on the young liberals in Florence "to decide on the major questions and to put forward their thoughts clearly and precisely", since "they don't know what they want and they have no program, so they will be defeated before they enter the fray". In particular, he urged them to make a choice, at least in one case, between free trade and protectionism, between adopting an aggressive foreign policy or otherwise, between accepting or not accepting parliamentarianism, between approving or otherwise the notion of a free church in a free state (considered by Pareto to be the "single thing that has worked in Italy"), between the regulation of strikes or otherwise, between management of services by the state or otherwise.

In April 1901 Pareto declared that in that historic moment, the duty of liberals was to combat "intolerance, whatever its provenance"; if not, "their party will have lost its raison d'être, and they themselves will have destroyed it".[30]

Furthermore,[31] although he acknowledged that "politics is simply a compromise and a party that wishes to govern has to compromise", he observed that in Italy too the liberal party had practically disappeared "by dint of compromising". To him its disappearance seemed inevitable because "how can a party attract supporters if it spends its time belying its principles through its actions?". From that time on, the notion of the self-defeating character of the liberals was a constant refrain.

Thus, in June 1903,[32] Pareto noted that over the past ten years or more, "the English liberals have expended all their energies in curtailing the country's freedoms". Then, in October 1903,[33] harking back to the liberalism of his youth, he recognised that it had been founded on the "idea, which was [apparently] logical, but which in reality was mistaken" of combatting the opponents of that policy (designated generically as A) which he considered to be "productive for society". This idea, in fact, would have been "correct [only if] the choice had [been] between A and non-A alone", thus implying that combatting non-A would definitely have benefitted A and society. Instead, combatting protectionism, for example, would be of benefit to society only if the result of a victory over protectionism meant the success of free trade. However, the overall effect on society would be uncertain if the defeat of protectionism led to the creation of a "socialist parliament" like the Paris commune, as was the case in France after Napoleon III. Pareto was forced to confess his error when "[his] study of history revealed that in the nineteenth century those who were fighting for liberty were, unknowingly and unintentionally, its fiercest enemies", as was still the situation in France and in Switzerland[34] "[where] it was clear that [the liberals] were beneficial only to those who desired the disappearance of the last vestiges of liberty".

However, if the liberals were detrimental to liberty, albeit involuntarily,[35] then neither did the parties of the left defend it. The populist parties may have invoked liberty when they were in opposition but they denied it when they were in power;[36] in point of fact, "liberty is nobody's property and cannot be said to belong to the extreme democrats [because] liberty is a luxury which can be appreciated only after life's more immediate needs have been satisfied".[37]

3.3 Theoretical Considerations on the Causes of the Liberalism's Defeat

In his *Systèmes Socialistes*, Pareto dwells, among other things, on the reasons for the failure of liberalism.[38]

The main reason why economic liberalism, in particular, never completely succeeded in establishing itself was that "it offers only justice and well-being for the greatest number, which is not enough", because in order to attract followers, it is necessary to offer them privileges:[39] thus, the success of the Cobden league, too, was "simply a triumph of certain interests over certain others".[40] Moreover, the liberals had lost also because they had thought they could "direct the masses through reason alone", whereas people can be spurred to action only "by engaging with their feelings and with their interests".[41] In Pareto's view, the doctrine of liberty (which requires "the indispensable adjunct and corrective of responsibility"[42]) would not be utopian could it at least induce people to stop "plundering each other through the offices of the law",[43] a change which might be simpler to achieve than persuading them to worry more about the public interest than their own private ones (being the essential prerequisite for the implementation of collectivism).[44] But in the end, liberal principles, even if "they can be intrinsically beneficial to society", evidently "must harbour something which is distasteful to existing human nature".[45]

Notes

1. See (Mornati 2018, particularly chapter II).
2. Pareto to Federigo Enriques, 26th December 1906, BPS-la (Banca Popolare di Sondrio: Vilfredo Pareto's letters archive).
3. Ibid.

4. This period has received extensive coverage in Italian contemporary political historiography. See, for example, Romanelli (1990) and Cammarano (1999).
5. Pareto to Francesco Papafava, 28th May 1899, see (Pareto 1989, p. 351).
6. See (Pareto 1899a, reprinted in Pareto 1987, p. 315).
7. On this topic, in March 1899, Pareto, while generally deploring the continuation of the Italian government's colonialist policy, had described as "throwing money away" the temporarily stalled Italian bid to purchase the Sa Mun bay in China, Pareto to Maffeo Pantaleoni, 12th March 1899, see (Pareto 1984a, p. 263).
8. See (Pareto 1899b, reprinted in Pareto 1987, pp. 317–318). A few months later Pareto remarked, in similarly bitter tones, that "people who in Italy (and also elsewhere) say that they stand for liberty, do not want to sacrifice money, time, friendships, or anything else" Pareto to Papafava, 13th February 1901, see (Pareto 1989, p. 375).
9. See (Pareto 1899c, reprinted in Pareto 1987, p. 321).
10. Pareto to Pantaleoni, 4th December 1899, see (Pareto 1984a, p. 284).
11. Some time later, Pareto added that he had witnessed the devotees of Dreyfus (at the beginning of the 1900s, in the wake of their victory) "using the same black arts about which they had protested against their adversaries", which had shown him that "while a few dupes like [him] attempted to follow their principles, the majority only looked after their own interests", Pareto to Alceste Antonucci, 7th December 1907, see (Pareto 1975, p. 615).
12. Consequently, Pareto deplored Pantaleoni's recent decision to enter politics, succeeding in being elected to parliament on the radical-socialist list in the general election of June 1900, Pareto to Arturo Linaker, 29th October 1900, see (Pareto 1975, p. 412).
13. Pareto to Papafava, 22nd June 1900, see (Pareto 1989, pp. 367–368).
14. Pareto to Pantaleoni, 18th July 1900, in Pareto, *Lettere a Pantaleoni (Letters to Pantaleoni) 1897–1906*, p. 320.
15. Pareto to Pantaleoni, 16th August 1900, ibid., p. 325.
16. Pareto to Armand Massip, 7th March 1901, BPS-la.
17. Pareto to Giuseppe Jona, 14th October 1903, BPS-la.
18. Pareto to Papafava, 12th June 1905, see (Pareto 1989, pp. 367–368).
19. Pareto to Carlo Placci, 4th January 1904, see (Pareto 1975, p. 513).
20. Pareto to Papafava, 29th October 1904, see (Pareto 1989, p. 441). In December 1903, he had claimed not to be an individualist because "this term belongs to the field of metaphysical doctrines which is foreign to me", Pareto to Julius Wolf, December 1903, BPS-la.
21. Pareto to Pantaleoni, 1st April 1905, see (Pareto 1984a, p. 442).
22. Pareto to Luigi Bodio, 27th December 1904, see (Pareto 2001, p. 157).

23. Pareto to Linaker, 29th October 1900, see (Pareto 1975, p. 400).
24. Pareto to Pantaleoni, 24th September 1900, see (Pareto 1984a, p. 341.
25. Pareto to Adrien Naville, 24th January 1900, see (Pareto 1975, p. 400).
26. Pareto to Linaker, 4th August 1901, see (Pareto 1975, p. 432). Similarly, Pareto to Pantaleoni, 3rd July 1902, see (Pareto 1984a, p. 405) and Pareto to Placci, 4th January 1904, see (Pareto 1975, p. 513).
27. Pareto to Papafava, 12th June 1905, see (Pareto 1989, pp. 445–446).
28. Pareto to Pantaleoni, 7th March 1907, see (Pareto 1984b, p. 17).
29. Pareto to the President of the Camillo Cavour Association, 4th February 1901, BPS-la.
30. See (Pareto 1901, p. 80).
31. Pareto to Gustave de Molinari, 10th June 1901, BPS-la.
32. See (Pareto 1903).
33. Pareto to Jona, 14th October 1903, op. cit.
34. Referring to the federal referendum of 15th March 1903 when 59.6% of the voters had approved the federal customs law of 10th October 1902, which was more protectionist than the 10th April 1891 law it replaced. Pareto concluded that in Switzerland "[direct] democracy is the worst enemy of liberty", Pareto to Tullio Martello, 17th February 1905, see (Pareto 1975, p. 538).
35. "Through examining and re-examining history", Pareto concluded that in general "people labour in order to bring themselves nearer to their desired objective P but instead, by virtue of their actions, they bring society nearer to a condition of Q which they would never ever have wished for", Pareto to Alceste Antonucci, 7th December 1907, see (Pareto 1975. p. 615).
36. See (Pareto 1905, reprinted in Pareto 1987, p. 405).
37. Ibid., p. 412.
38. As regards the work of the liberal economists, Pareto pronounced that "the scientific part is good ... it is the foundation of the modern science of economics", whereas "the metaphysical part is certainly worth no more than other similar flights of fancy", see (Pareto 1902, p. 46.) Pareto added further that "liberal economists have an excessively narrow standpoint, giving too much importance to the economics and not enough to the sociology", Pareto to Papafava, 12th July 1902, see (Pareto 1989, p. 404).
39. See (Pareto 1902, p. 92).
40. Ibid.
41. Ibid., pp. 66–67.
42. Ibid., p. 57.
43. Ibid., p. 56.
44. Ibid.
45. Ibid., p. 419.

Bibliography

Cammarano, Fulvio. 1999. *Storia politica dell'Italia liberale. L'età del liberalismo classico, 1861–1901 [Political history of liberal Italy. The age of classical liberalism, 1861–1901]*. Rome-Bari: Laterza.

Mornati, Fiorenzo. 2018. *An intellectual biography of Vilfredo Pareto, II, Illusions and delusions of liberty (1891–1898)*. London: Palgrave Macmillan.

Pareto, Vilfredo. 1899a. Perché siamo uniti? [Why are we united?]. *Critica sociale*, July, 16, pp. 148–149

———. 1899b. L'elezione del V collegio [The election in the 5th constituency]. *Il Secolo*, August 10–11.

———. 1899c. Liberali e socialisti [Liberals and socialists]. *Critica sociale*, September 1, pp. 215–216.

———. 1901. A propos de la loi sur les associations [Regarding the law on associations]. *Journal des économistes* LX (XLVI-1): 76–80.

———. 1902. *Les systèmes socialistes [Socialist systems]*, tome II. Paris: Giard et Brière.

———. 1903. L'éclipse de la liberté [The eclipse of liberty]. *Gazette de Lausanne*, June 8.

———. 1975. *Epistolario 1890–1923 [Correspondence, 1890–1923]*. Complete Works, tome XIX-1, ed. Giovanni Busino. Geneva: Droz.

———. 1984a. *Lettere a Maffeo Pantaleoni 1897–1906 [Letters to Maffeo Pantaleoni 1897–1906]*. Complete Works, tome XXVIII.II, ed. Gabriele De Rosa. Geneva: Droz.

———. 1984b. *Letters to Maffeo Pantaleoni 1907–1923 [Lettere a Maffeo Pantaleoni 1897–1906]*. Complete Works, tome XXVIII.III, ed. Gabriele De Rosa. Geneva: Droz.

———. 1987. *Écrits politiques. Reazione, Libertà, Fascismo, 1896–1923 [Political writings. Reaction, liberty, fascism, 1896–1923]*. Complete Works, tome XVIII, ed. Giovanni Busino. Geneva: Droz.

———. 1989. *Lettres et Correspondances [Letters and correspondence]*. Complete Works, tome XXX, ed. Giovanni Busino. Geneva: Droz.

———. 2001. *Nouvelles Lettres 1870–1923 [New Letters 1870–1923]*. Complete Works, tome XXXI, ed. Fiorenzo Mornati. Geneva: Droz.

Romanelli, Raffaele. 1990. *L'Italia liberale [Liberal Italy]*. Bologna: Il Mulino.

CHAPTER 4

The Advance of Socialism and the Obstacles Impeding It

Until the eve of the First World War, Pareto viewed socialism as the winning political force and from the mid-1880s he devoted his meticulous and often sympathetic scrutiny to it.[1] In this chapter, drawing principally but not exclusively on his *Systèmes Socialistes*,[2] we will characterise the definitive conception Pareto arrived at in relation to socialism and track the continuing process of observation he dedicated to its apparently unstoppable advance in the period from the beginning of the century until the outbreak of the First World War.[3]

4.1 Socialism: Definitions and Common Principles of the Various Theories

According to Pareto, a clear criterion "for the classification of social systems" is to be found in the extent to which they recognise private property.[4] Thus, the situation where no private property exists corresponds to communism[5]; a context where ownership of the means of production is abolished while private ownership of what is produced is retained characterises "modern socialist [and] government monopoly systems",[6] while state socialism refers to the situation where private ownership of the means of production is maintained but ownership of what is produced is abolished.[7]

F. Mornati, *Vilfredo Pareto: An Intellectual Biography Volume III*,
Palgrave Studies in the History of Economic Thought,
https://doi.org/10.1007/978-3-030-57757-5_4

Theoreticians' ideas on socialist systems had in common a claimed ability (which, in fact, was never proven) to place the government in the hands of "the best and the most competent"[8] through the agency of their proposed mechanisms, without considering that "at the present time the only way we have of making a selection [among rival candidates] is to put them in competition with each other and see what they can do".[9]

This aside, in Pareto's view, socialist theoretical systems can be divided into differing religious, metaphysical and scientific categories, the first referring to systems which are based on certain religious sentiments,[10] the second aiming to influence human behaviour on the basis of non-empirical principles[11] and the last striving to promote human happiness, while at least attempting to make some reference to experience and logic.[12]

4.2 Religious and Metaphysical Systems

The type of system which, prior to the time of writing the book, had had the most numerous practical manifestations was the religious one, confirming that the transformation of the human character indispensable for modifying the social order can come about only by harnessing the force of religious sentiment.[13] Religious socialist systems which had actually been put into practice had been made possible only by carrying altruistic sentiments to an extreme. However, this type of approach is to be found only in small and highly motivated *élite* groups, implying that such communities had had to revert to the (non-altruistic) social norm when they had attracted a larger number of adherents.[14]

Pareto remarked that it is rather the metaphysical and scientific systems which "acquire a greater importance" when seen from a subjective point of view, because people like to be guided by metaphysical or scientific considerations rather than by religious ones, even in relation to socialist issues.[15] According to Pareto, Plato himself was not aiming directly at the achievement of human happiness but rather at the creation of an ideal state (i.e. one in which each individual performs the role assigned to him) which would be of benefit to the populace,[16] so that "Plato's socialism is principally an ethical socialism".[17]

4.3 The Scientific Systems, Particularly the Economic Component of Marx's Thinking

Among scientific systems, Pareto included and gave critical assessments for those of Thomas More, Charles Fourier and Pierre-Joseph Proudhon,[18] but his principal attention was reserved for the writings of Marx.

First, he pointed out that Marx's most dramatic prognostications had yet to come true, adding "economic crises have become milder, poverty has declined, the middle class has not disappeared, small-scale agriculture and industry persists and thrives, no increase in the concentration of wealth has been observed".[19] The predicted trend towards a reduction in profit margins was contradicted by the protracted rise in profits associated with the industrial revolution, in spite of the simultaneous marked increase in constant capital in relation to variable capital.[20]

The theory of labour value which occupies such an important place in Marx's economic thinking and which was clearly borrowed by Ricardo can, in turn, only be valid if we postulate that all the other values that it depends on are constant.[21] The main defect in Marx's economic theory is precisely that of attempting to "pass off this hypothesis, which is not only legitimate but also necessary, as reality".[22]

From the "logical and experiential refutation of Marx's theories", specifically the proposition that "the origin of profit is when labour produces more than is needed for its own maintenance",[23] it does not follow, however, that we have either "the right to conclude that collectivism would not be conducive to society's well-being"[24] or that "the influence that Marx's economic theory can exert to lead people in a certain direction" can be denied.[25] In Pareto's opinion this influence is derived from "Marx's powerful dialectic [which] demands the respect merited by any adversary equipped with powers which are out of the ordinary".[26]

Shortly before this, Pareto had made a distinction between scientific Marxists ("who sincerely seek the truth" and attempt to reconcile it "with Marx's book"[27]), orthodox Marxists ("who perfectly understand the importance of capital, but they desire it to be collective instead of belonging to private individuals"[28]) and the acolytes of "vulgar socialism", particularly the radical socialists ("who favour the destruction of private capital without giving any thought to the establishment of collective capital",[29] which "must necessarily lead sooner or later to the impoverishment of the nation"[30]).

Pareto devoted particular attention to the work of the young Italian Marxists of the time. Commenting on Benedetto Croce's recent paper,[31] Pareto[32] remarked once again that "Marx's theory suffers from the grave defect of never agreeing with the facts". In fact, alluding to the theory of labour value, Pareto pointed out to Croce that in a flower catalogue he had found that the prices had almost doubled from 1897 to 1898 while the quantity of work had decreased with the passing of time "because one learns with experience". Furthermore, in reference to Croce's interpretation whereby the theory of labour value is valid only for an ideal society composed entirely of workers producing exclusively through their own labour,[33] Pareto added that for this to be true, it would be necessary to add the further absurd proviso that, in this society, people's tastes must all be identical.

Croce's next work[34] offered Pareto the opportunity to make further critical observations regarding Marxist economic theory. Firstly, the decline in business activities[35] where variable capital was, by comparison to competing enterprises,[36] proportionally greater than constant capital again belied Marx's theory whereby "profits are higher to the degree that variable capital is proportionally greater than constant capital". Secondly, the production coefficients (i.e. the organic makeup of capital) which Marx viewed as determined "technically" are in fact determined "economically" by reference to the interest rate and to salaries, with the ratio of constant to variable capital decreasing (increasing) with increases (decreases) in the interest rate and/or with decreases (increases) in salaries, and where the interest rate and salaries depend in turn on the organic makeup of the capital.[37]

He also added, firstly, that Marx did not define value, oscillating between the value-in-exchange and the value-in-use; that he was erroneously convinced that he had identified all the characteristics of a phenomenon and hence, wrongly, that having shown that if a given quality does not depend on some of the characteristics he had identified, then it must depend on others he had distinguished in his partial analysis; and, finally, that he had completely neglected to perform the "experimental verification" of his supposed laws which, instead, was indispensable.[38]

Pareto ended by conceding that "in practice socialism, which is a faith, needs a sacred book … [containing] its mysteries", a function which was duly performed by Marx's *Das Kapital*, while adding that, nevertheless, he was "surprised to see scientists [such as Croce] following the same path".[39]

In a review of Arturo Labriola's first volume on Marx's theory of value, thanks to which "Marx's ideas become clearer",[40] Pareto took the opportunity to repeat that, firstly, Marx's interpretation of the economic system as embodied in the sequence money-commodity-money "is completely mistaken, [as] in our modern societies money is involved in only a minute fraction of exchanges"; secondly, there is no reason to conclude, as Marx does, that "in capitalist societies the value of capital must continue to increase"; thirdly, in general, "Marx's works are not only full of factual errors, they also constitute models of false reasoning". He therefore advised Marxists to devote themselves exclusively to the study of the facts, instead of continuing their "subtle exegesis" of a "language vague and obscure". Thus, in Pareto's view, Vincenzo Giuffrida, with his book on the third volume of Marx's *Kapital*, had performed a major service for Marxists by encapsulating, in only 130 pages, the "arguments of a perplexing tedium and obscurity" contained in Marx's works: in his review of Giuffrida, Pareto again rejects Marx's claim that value is determined only by the cost of production and not by "all the conditions" which influence "economic phenomena".[41]

4.4 The Scientific Systems: The Sociological Component of Marx's Thought

In Pareto's opinion, on the other hand, "the sociological elements in Marx, such as the notion of class struggle, often correspond to reality".[42]

Pareto, after repeating that, contrary to the popular interpretation of the materialist conception of history whereby "economic conditions determine other social phenomena", there is simply an interdependence between economic and other social phenomena,[43] stressed that there is also an interpretation of the materialist perspective which "sees historical events between which it is necessary to discover the relationships" and hence "it has all the hallmarks of a scientific theory".[44]

Similarly, the concept of class struggle has, firstly, a popular interpretation whereby society is divided into two classes, proletarians and capitalists, with the former ending up by violently destroying the latter,[45] as well as, secondly, an interpretation derived from the materialist conception which envisages "a great number of classes" among which there are "infinite" forms of struggle[46] and where each class must count only on itself to defend its own interests.[47]

In any case, having recognised the existence of class struggle, Pareto remarked that in order to assess it "equitably",[48] it is necessary to realise that often the only choice in social relations is between allowing oneself to be despoiled and demanding one's share of the spoils, implying that moderate parties will disappear and "only the hardline parties will remain in contention".[49] Pareto then underlined that "we are a long way from a simple struggle between two classes", since both the bourgeoisie and the proletariat are divided internally, the former between industrial and agricultural protectionists and the latter between socialists and anarchists (with the socialists being further subdivided into hardliners and moderates) as well as between unionised and non-unionised workers.[50]

Pareto repeated that there was the further "question" regarding the methods to be adopted in the class struggle, as well as the link between these and the well-being of society.[51] Here Pareto, after observing that in this struggle direct recourse to violence decreases in frequency with the increasing sizes of social groups[52] and that there can also be "an intelligent class struggle", as in the case of the English Trade Unions, who "do not minimally seek the ruin [of the entrepreneurs but] on the contrary, want them to make plenty of money so as to obtain their share of the profits",[53] added that in any case the time had not yet come to expect the complete disappearance of violence from social relations.[54] Further, on the basis of the general principle that "from the point of view of social organisation, conflict [between a number of social forces] seems a lesser evil than if one force alone could act without any counterweight",[55] he considered it useful for society when entrepreneurs join together "to form an appropriate opposition" to workers' demands (as in the case of the American Trusts).[56]

Pareto concluded that even if collectivism took over,[57] class struggle would continue notwithstanding, representing essentially "the fight for survival",[58] perhaps no longer between proletarians and the bourgeoisie but "between intellectuals and non-intellectuals, between different categories of politicians, between the latter and their citizens, between innovators and conservatives".[59]

4.5 Socialism Seen According to the Criteria of Pure Economics

In July 1899, Pareto told his friend, the philosopher Adrien Naville from Geneva, that a corporate social regime, equivalent to state socialism, would not be able to feed "the [whole] existing population" because it would suppress the mechanism of competition, which represented the sole guarantee of productive efficiency, facing "the populace with the harsh necessity of modifying its habits or being ruined".[60]

In the same period, with reference to a recent book on the organisation of the future socialist society by his friend and ex-colleague Georges Renard (a communard in long-term exile in Lausanne), he commented that it was an example of the type of non-Marxist socialist doctrine which "tends to become a sort of religion", containing largely arbitrary propositions "from which a host of conclusions are drawn".[61] In particular, Pareto noted and criticised the following claims advanced by Renard as being unproven: firstly, the idea that the collectivisation of land ownership would be able to provide the populace with an assortment of spacious dwellings, on much more favourable terms than those provided by the individualist society; and secondly, the notion that it would be sufficient to "reverse the positions of governors and governed [for] further abuses and oppression to become impossible".[62] He also doubted whether the maximum of well-being for society could be achieved through Renard's proposal to reward workers for their labour on the basis of the proportion between the number of hours of labour needed by society in order to produce a given quantity of a commodity and the number of workers spontaneously making themselves available, as it might always be in a socialist government's interest to modify the quantity of work associated with each commodity in order to obtain the broadest selection of commodities possible, thus arriving "by dint of fine-tuning, at the same solution as free competition would yield".[63] In general terms, Pareto noted that "in modifying the manufacturing coefficients [of free trade], society, taken as a whole, will suffer [since] the same amounts of work and of capital will yield fewer goods". Thus, if the "working class" succeeded in imposing production coefficients involving greater quantities of labour than the optimum, very rarely would they derive a clear benefit from the step because, given that they constitute "the overwhelming majority of the population", they would pay a large part of the price for the damage caused to society's material well-being by the adoption of non-optimal production coefficients.[64]

Pareto then observed that various writers also aimed to show that "at one time property was collective", both in an attempt to seek compensation for the population concerned and also in order to be able to assert that "collective production is possible, because it has existed in the past".[65] After objecting that it would be impossible to identify who should pay this putative compensation[66] and that "it remains to be shown that people were happier under the regime of collective land ownership than under that of private property", Pareto added "one often encounters evidence" showing that "a collective farming regime is perfectly compatible with a social organisation which encompasses poor and rich".[67] Additionally, it was far from proven that collectivisation of agriculture yielded "the highest output".[68]

Following these critical observations on various blueprints for socialist systems, in his *Manuale di economia politica* Pareto once again advanced his idea for the optimal form of collectivism.

Having characterised "collectivist society [as] having the aim of deriving the maximum of ophelimity for its members"[69] and considering the "problem [of] how to share out the assets owned or produced by the society among its members"[70] to have been resolved, Pareto underlined that collectivism needed to address the problem of "how to produce economic goods so that the members of the society derive the maximum of ophelimity".[71] To achieve this, collectivism would need to "determine the production coefficients in the same manner as free competition does" and "equalise the various forms of yield on capital", as happens automatically under free competition.[72] Hence free competition systems and collectivist systems, thus defined, differ only in the distribution of income which, under free competition, "emerges from all the historical and economic circumstances through which society developed", while in the case of collectivism it is "fixed by reference to certain ethical and social principles".[73] Pareto further added that, possibly, income distribution could be modified by exchanging private property for collectivism, but to him "it seems unlikely that no hierarchy would persist" in terms of earnings.[74]

Next, Pareto noted that individualistic societies are able to reach only a point of intersection between the line of exchange and the line of complete transformation, whereas the maximum of ophelimity is given by the point of tangency between this and the curve of indifference,[75] implying that "there must be an enormous loss of ophelimity".[76] In other words, in order to arrive at the point of maximum ophelimity, a producer would have to receive payment in advance from his customers to cover his "general expenses" and then sell back "the commodities at cost price".[77] Pareto, "except in special cases, cannot see how a private company" could do this, whereas the socialist state "can charge the consumers of a commodity for the general expenses involved in the production of that commodity in the form of a tax and then sell it at cost price".[78] Pareto observed that "the benefit that society would derive might be sufficient to compensate the inevitable loss involved in this sort of production … [on condition that] the aim was exclusively to achieve the maximum of productive ophelimity and not to generate monopoly profits for the workforce or to pursue humanitarian ideals".[79] On this topic, Pareto noted that "the majority of trade unions", defending the interests of their members, reduced the number of apprentices and also obliged employers to hire only workers who have union membership, which they themselves made as difficult as possible to obtain.[80] Thus, for example, in the building industry, there would be a small number of construction workers with very high wages and very few people "with enough money who could afford the luxury of building".[81] In the end, Pareto thought that even if "privileges" can emerge in any society, nevertheless "sooner or later, revolutions, violent or peaceful, come to destroy [them]".[82]

A second characteristic is that "the socialist state can give the enjoyment of the rent deriving from the production of a commodity to the consumers of that commodity".[83]

Nevertheless, since production under a collectivist state "would be controlled by employees of that state", which might provoke "higher costs and less satisfactory work" compared to production overseen by "entrepreneurs and landowners in a state with private ownership",[84] counterbalancing the advantages of collectivism mentioned above. Hence, Pareto observed that "pure economics does not supply us with truly decisive criteria for choosing between a system of private property or free competition and a socialist system".[85]

An algebraic analysis of the decisions of a "collective wishing to regulate production in the most beneficial way possible for its members" starts out from the supposition that each of them is assigned "on the basis of criteria deemed appropriate", an initial allocation of commodities X, Y, … and A, B, ….[86] Simplifying the notation, for each individual there will be the following variations: in the quantities produced of commodities X, Y, …; in the quantities used (for production purposes) of commodities A, B, … and consequently in individual ophelimity[87]:

$$d\Phi_1 = \varphi_{1x} dx_1 + \varphi_{1a} da_1 + \varphi_{1b} db_1 + \ldots$$
$$d\Phi_2 = \varphi_{2x} dx_2 + \varphi_{2a} da_2 + \varphi_2 db_2 + \ldots$$
$$\smile$$

which, in the light of

$$\varphi_{1a} = \frac{\varphi_{1x}}{p_x} = \frac{\varphi_{1b}}{p_b} = \ldots; \varphi_{2a} = \frac{\varphi_{2x}}{p_x} = \frac{\varphi_{2b}}{p_b} = \ldots; \ldots$$

or

$$p_x = \left(\frac{\varphi_{1x}}{\varphi_{1a}}\right), p_b = \left(\frac{\varphi_{1b}}{\varphi_{1a}}\right) \ldots; p_x = \left(\frac{\varphi_{2x}}{\varphi_{2a}}\right), p_b = \left(\frac{\varphi_{2b}}{\varphi_{2a}}\right) \ldots;$$

or alternatively

$$p_x = \left(\frac{\varphi_{1x}}{\varphi_{1a}}\right) = \left(\frac{\varphi_{2x}}{\varphi_{2a}}\right) = \ldots; p_b = \left(\frac{\varphi_{1b}}{\varphi_{1a}}\right) = \left(\frac{\varphi_{2b}}{\varphi_{2a}}\right) = \ldots; \ldots$$

become

$$\frac{d\Phi_1}{\varphi_{1a}} = p_x dx_1 + da_1 + p_b db_1 + \ldots$$
$$\frac{d\Phi_2}{\varphi_{2a}} = p_x dx_2 + da_2 + p_b db_2 + \ldots$$
$$\smile$$

summing which we obtain

$$\left(\frac{d\Phi_1}{\varphi_{1a}}\right) + \left(\frac{d\Phi_2}{\varphi_{2a}}\right) + \ldots = p_x dX + dA + p_d dB + \ldots$$

Since the functions

$$A = F(X,Y,\ldots); B = G(X,Y,\ldots);\ldots$$

apply, from which the following equations can be derived.[88]

$$dA = \left(\frac{\partial F}{\partial x}\right)dx = -a_x dX; dB = \left(\frac{\partial G}{\partial x}\right)dx = -b_x dX;\ldots$$

the previously specified

$$\left(\frac{d\Phi_1}{\varphi_{1a}}\right) + \left(\frac{d\Phi_2}{\varphi_{2a}}\right) + \ldots = p_x dX + dA + p_b dB + \ldots$$

becomes

$$\left(\frac{d\Phi_1}{\varphi_{1a}}\right) + \left(\frac{d\Phi_2}{\varphi_{2a}}\right) + \ldots = p_x dX - a_x dX - b_x p_b dX + \ldots = (p_x - a_x - p_b b_x - \ldots)dX + \ldots$$

where the first element made equal to zero, or

$$\left(\frac{d\Phi_1}{\varphi_{1a}}\right) + \left(\frac{d\Phi_2}{\varphi_{2a}}\right) + \ldots = 0$$

yields an algebraic representation of the definition of the maximum of collective ophelimity as "a position [from which] it is impossible to diverge even minimally without benefitting or harming all the members of the group", that is, a position "any minute deviation [from which has] the inevitable effect of benefitting some members of the group and harming others".

Accordingly, remembering that φ_{1a}, φ_{2a}, … are positive and homogeneous with regard to Φ_1, Φ_2, …, which authorises us to divide $d\Phi_1$ by $d\varphi_{1a}$ … obtaining $(d\Phi_1/\varphi_{1a})$, $(d\Phi_2/\varphi_{2a})$, … which, in turn, are homogeneous and hence comparable (in that they are all made up of quantities of commodity A), the equation is satisfied if some $d\Phi_1$, $d\Phi_2$, … are positive and others negative.

Thus, when the maximum collective ophelimity is attained

$$\left(\frac{d\Phi_1}{\varphi_{1a}}\right)+\left(\frac{d\Phi_2}{\varphi_{2a}}\right)+\ldots=\left(p_x-a_x-p_b b_x-\ldots\right)dX+\ldots$$

becomes

$$0=p_x-a_x-p_b b_x-\ldots\text{or } p_x=a_x+p_b b_x+\ldots$$

from which we can conclude that "to attain equilibrium together with maximum ophelimity it is necessary not only for the total cost of production of the commodity to be equal to the total expenditure for consumption[89] but also for the cost of production of the last particle to be equal to the selling price of that last particle".[90]

4.6 Monitoring the Rise of Socialism in Practice

On the other hand, the bourgeoisie, so reviled by Pareto, appeared to do everything they could to facilitate the socialists' explicit aim of "dispossessing them",[91] an objective which was shared by all the various socialist factions.[92]

In January 1900, having stated that his recent article *The socialist tide (La marée socialiste)* was directed not "against the communists but against the much more dangerous state socialists", Pareto expressed his view that "we are surrounded by socialism" citing the examples of Geneva, where the socialists "are in power through the offices of the radicals",[93] and, more particularly, of France, where he feared that the socialists "are preparing to launch a revolution worse than the Terror or the Paris Commune".[94]

Soon afterwards, Pareto considered it "inevitable", due not so much to the strength of the socialists as to the culpable weakness of the bourgeoisie, that there would be "an economic revolution [which will again extend] over Europe the darkness of the middle ages".[95] Pareto also anticipated that if the bourgeoisie really were defeated, it would not be by the hardline socialists but by the moderates, because the extremists were capable of spreading a new doctrine effectively, but "once their work is done, they must necessarily step back and leave their place to the moderates".[96]

The moderate socialists, for their part, who "live as parasites in this bourgeois society which they say they wish to destroy but which is necessary to them",[97] would not be "less rapacious" than the hardliners[98] and in the end would simply embody "the advent of a new aristocracy".[99] In this regard, Pareto recollected that as had "happened in the past, the leaders during the period of transition were almost always (possibly always) drawn from among the members of the declining aristocracy".[100] Thus, the socialist leaders originated from the bourgeoisie, which "does not mean that when the revolution they are preparing comes, the bulk of the armed forces will not be supplied by the working class". While judging them to have conserved more vigour than their English or French counterparts, Pareto "could not say whether the Italian bourgeoisie still have enough vitality to save themselves from decline".[101] However, shortly before the Italian general election of 6th and 13th November, Pareto anticipated that "the revolutionary socialist party will probably win because it has no scruples about resorting to force, which since the dawn of time has been the only factor which can deliver, protect and maintain victory",[102] even if force "needs to be a means and not an end".[103]

Notwithstanding the Italian socialist party's disappointing electoral results,[104] Pareto continued to scrutinise the revolutionary Trade Union movement,[105] which, although "it preaches violence" like the anarchists, still remains "a collectivist party" and Pareto duly credited it with the capacity, which the anarchists lacked, to "be able to organise a society".[106] In Pareto's opinion, revolutionary Trade Unionism represented a return "to Marx's ideas on the class struggle becoming implacable, ferocious, savage" and also emerged from "appetites" aroused by "decadent humanitarian socialism" but which it had been unable to satisfy.[107] Further, referring in September 1906 to the unions which were being set up within the public administration in Italy, Pareto avowed that "we are heading rapidly for complete anarchy", while at the same time adding "so much the better: new people will arrive who will not be afraid of bloodshed and who will bring back order to the country".[108] Lastly, in April 1908, during the strikes in Parma which the revolutionary Trade Unionists had organised against the local landowners, Pareto had his "fingers crossed for the Trade Unionists to win and banish those idiotic bourgeois who are allergic to bloodshed [, ignoring the fact that] the earth has always belonged to those able to protect it by force".[109]

Pareto's last observations on the socialism of the *Belle Époque* date to the time of the outbreak of the First World War. He remarked that socialism "is unable to withstand nationalism or imperialism" because it was pervaded with an instinct for scheming, as was also shown by the fact that "a large number of socialists change their faith and, under various pretexts, join up with the nationalists and the imperialists".[110]

4.7 Nationalism

As we have already seen, as of the beginning of the century, Pareto viewed the European political battle[111] as a conflict between "the extremist parties" (i.e. socialism) and "the traditional religions and nationalism[112] or imperialism".[113]

As early as February 1902,[114] he was of the view that even if "the socialist tide continues to rise", nevertheless "in the end the reaction will come, brutal, implacable", adding that "no man can negate the fact that the more forceful the action, the more forceful the corresponding reaction". On the other hand, the necessary re-grouping of the anti-socialist forces was slow in coming about,[115] although Pareto considered that "the bourgeoisie should still have enough strength to oppose the foe effectively",[116] were it able to summon the determination.

In December 1903, Pareto nevertheless observed that one factor which spoke in favour of nascent Italian nationalism was its capacity to resist socialism by counterposing "another faith and another religion".[117] On the other hand, the bellicose and imperialistic schemes of Italian nationalism appeared to him to be unrealistic due to the unsustainable cost of military conflict (with the example of the recent Boer war freshly in mind[118]). Hence, these schemes may "occupy the forefront of our mind but should not be too openly discussed, in the same way as sacred relics".[119]

Despite his many perplexities, in February 1904, he lauded the contributors to the Florentine review "Il Regno"[120] because they were the only ones "who have the courage to combat the socialists", helping, through their anti-socialist nationalism, to re-establish the social equilibrium and to defend their class.[121] Once more shortly before the world war, Pareto credited nationalism with the merit of satisfying the need which "people have for an ideal"[122] and of returning "the pendulum of contemporary ideas from a materialistic-humanitarian extreme to an alternative position which can be considered more fruitful, at least in terms of being less extravagant".[123]

4.8 A Great War as the Only Possible Means to Bring About an Anti-socialist Realignment of Society

It was early in 1904 that Pareto brought to a close his last area of political activism when he conveyed his "best and warmest congratulations" to the editor of the *Gazette de Lausanne*, Edouard Secrétan, for having expressed "perfectly correct ideas" regarding the illusory nature of antimilitarism.[124125]

What is more, "war can change many things and, for a while, can steer our societies away from the precipice towards which they were heading".[126] More specifically, Pareto anticipated that "if there is a great European war, socialism will be set back by at least half a century and the bourgeoisie will be safe for that period".[127] In the summer of 1905, Pareto, while continuing to oppose international "aggression",[128] considered that "a European war is highly probable, firstly because of the growing and increasingly intense economic rivalry between Germany and England and secondly because the debacle in Russia and the democratic disintegration of the French army have modified the equilibrium which has underpinned the peace that we currently enjoy".[129] However, just before the war started, in March 1914 when he was still convinced that "there will be no war now, the financiers want military expenditure, but not war"[130] and he was no longer sure that "this war will be negative for socialism" because when it finished the defeated might turn to socialism (as in the case of the 1871 Paris Commune) as might the victors (like German socialism in the wake of the Franco-Prussian war).[131]

As for Italy's prospects for war, in 1905 Pareto opined that the country was "absolutely unprepared materially and, worse, morally".[132] However, if it continued to be an ally of Germany, "it will be on the side of the probable winner and hence will enjoy the fruits of victory", while if it went over to the other side, "very probably it will find itself among the defeated".[133] Then, at the time of the Bosnian crisis of 1908–1909, he noted that "Italy is becoming nationalistic", fearing that "it would all finish badly" in the light of the parallels between the current political mood in Italy and the mood "in France, prior to the war of 1870".[134] However, in reference to the war in Libya, of which he had immediately announced his approval,[135] Pareto claimed that Italy "could possibly not have avoided it without great danger",[136] since it was simply another consequence of the same "combination of interest and sentiment which, for the last century at least, have led to the colonial wars of all the major European nations".[137]

Notes

1. See (Mornati 2018a, §6.8) and Mornati (2018b, §2.8 and chapter IX).
2. Which Pareto claimed to have "written for purely scientific ends, eschewing any attempt at propaganda", Pareto to Giovanni Vailati, 30th June 1901, see (Pareto 1975a, p. 430). A few years later, Pareto stated that he had written the *Systèmes Socialistes* not for those wishing to enter the political fray [but] for "those who witness the comedy of politics and laugh", Pareto to Francesco Papafava, 28th May 1899, see (Pareto 1989, p. 460).
3. There is also an extensive bibliography on the history of Italian socialism in this period during which Pareto focused most of his attention on this political party. See, for example, (Ciuffoletti et al. 1992).
4. See (Pareto 1901a, p. 110).
5. Ibid.
6. Ibid., p. 111. These were endorsed by the most hardline socialists, ibid., p. 113.
7. Ibid., p. 111. In Pareto's view, state socialism garnered the support of "the landed classes [who believed they saw in it] a sort of insurance policy against revolutionary socialism [and also] by [humanitarians] who couldn't stomach the prospect of the revolutionary form of socialism" and was formulated as part of a legitimate attempt to resolve social problems by people who derived their theories, however, "from certain a priori principles which are more or less intelligible", instead of "founding them on real-world interconnections", ibid., p. 102. State socialism was endorsed by socialists of the more compromising sort too, who admitted the existence of private enterprise, but exclusively with a view to "squeezing as much as possible out of it", ibid., p. 113. He likewise noted that "the Marxists" who rejected state socialism acted "logically as … the majority of the provisions [of this type of socialism] are, arguably, infelicitous borrowings from the principles of Marxism", Pareto to Giovanni Borelli, 9th June 1901, BPS-la (Banca Popolare di Sondrio-Vilfredo Pareto letters archive).
8. See (Pareto 1901a, p. 269).
9. Ibid., p. 276.
10. Ibid., p. 264.
11. Ibid., p. 266 and (Pareto 1902a, chapters VIII, IX).
12. See (Pareto 1901a, p. 266) and (Pareto 1902a, chapter X).
13. See (Pareto 1901a, p. 266).

14. Ibid., p. 221.
15. Ibid., p. 268.
16. See (Pareto 1902a, p. 14).
17. Ibid., p. 8.
18. Ibid., chapter XI.
19. Ibid., p. 327.
20. Ibid., pp. 361–363.
21. Ibid., pp. 345–346.
22. Ibid., pp. 337, 349. However, according to (Pareto 1906a, chapter VII, §117), Marx's theory of labour value is simply "the expression of the feeling of discomfort which members of the new aristocracy experience when they are obliged to frequent the lower orders", and as they rise up the social scale their "feelings" change, transforming this theory into a "museum" piece.
23. See (Pareto 1902a, p. 330).
24. Ibid., p. 379.
25. Ibid., p. 380.
26. Ibid., p. 382.
27. See (Pareto 1900a, p. 163).
28. Ibid.
29. Ibid., 163–164.
30. See (Pareto 1901a, pp. 157–158). Undoubtedly, since the maximum for collective well-being derives from a relatively unchangeable combination of land, mobile and personal assets, any modification to any of these elements will gravely prejudice social well-being, ibid., p. 158. Pareto considered the decadence of ancient Athens and of the Roman empire to be illustrations of the effect of a reduction in mobile assets due to their squandering by state socialism, which had not been compensated by "increases in the productivity of industry and of agriculture", ibid., p. 161.
31. See (Croce 1899a).
32. Pareto to Benedetto Croce, 18th June 1899, see (Pareto 1975a, p. 384).
33. See (Croce 1898).
34. See (Croce 1899b). After repeating that, according to Marx, the rate of profitability depends on the ratio of variable to total capital (corresponding to the sum of variable and constant capital), Croce noted that, with no variations in production, technical progress will give rise to a reduction in total capital through an equiproportional reduction in constant and in variable capital (whence the proportional reduction in labour expended). It thus follows that the profit margin will remain unchanged, contrary to Marx's well-known thesis, while total profits alone will decline (being the result of addition of the reduced total profit and the rate of profitability which itself is unchanged).

35. For example, navigation by sail.
36. For example, navigation by steam.
37. Pareto to Croce, 18th June 1899, see (Pareto 1975a, p. 385).
38. Pareto to Croce, 7th July 1899, see (Pareto 1975a, 390).
39. Pareto to Croce, 18th June 1899, see (Pareto 1975a, p. 385). However, at that time he still believed that "the socialists are wrong in the economic sphere because they do not study it, but elsewhere they are right", Pareto to Teodoro Moneta, 18th November 1899, see (Pareto 2001, p. 123).
40. See (Pareto 1899a, reprinted in Pareto 1966, pp. 110–111).
41. See (Pareto 1899b, reprinted in Pareto 1966, pp. 112–114).
42. See (Pareto 1902a, p. 328).
43. Ibid., pp. 387–388.
44. Ibid., p. 390.
45. Ibid., pp. 402–403.
46. Ibid., pp. 404–405.
47. Ibid., pp. 408–409.
48. Ibid., p. 419.
49. Ibid., p. 419–420.
50. Ibid., p. 420.
51. Ibid., p. 409.
52. Ibid. On the other hand, the use of violence is better tolerated when the group involved is large.
53. Ibid., p. 436.
54. Ibid., p. 413.
55. Ibid., p. 446.
56. Ibid., pp. 444–445.
57. Ibid., p. 455.
58. Ibid., p. 454.
59. Ibid., p. 455.
60. Pareto to Adrien Naville, 1st July 1899, see (Pareto 1975a, pp. 388–389).
61. See (Pareto 1899c, reprinted in Pareto 1966, p. 141).
62. Ibid., pp. 141–142.
63. Ibid., pp. 143–144.
64. Pareto to Papafava, 27th April 1901, see (Pareto 1989, p. 378).
65. See (Pareto 1901a, p. 184).
66. Ibid.
67. Ibid., p. 185.
68. Ibid., p. 186.
69. See (Pareto 1906a, chapter VI, §52).
70. Ibid., chapter VI, §53.
71. Ibid.
72. Ibid., chapter VI, §55.

73. Ibid., chapter VI, §56.
74. Ibid., chapter VII, §23.
75. Ibid., chapter VI, §48.
76. Ibid., chapter VI, §44.
77. Ibid., chapter VI, §58.
78. Ibid., chapter VI, §58.
79. Ibid., chapter VI, §48.
80. See (Pareto 1902b, reprinted in Pareto 1966, p. 205).
81. Ibid., p. 206.
82. Ibid.
83. See (Pareto 1906a, chapter VI, §59).
84. Ibid., chapter VI, §60.
85. Ibid., chapter VI, §61.
86. Ibid., appendix, §43.
87. Ibid., appendix, §45.
88. Bearing in mind, firstly, that with increases in the quantity produced of a commodity the remaining quantity of the capital used for production diminishes and, secondly, that the production coefficients are positive.
89. Ibid., appendix, §36.
90. Ibid., appendix, §41.
91. See (Pareto 1899d, reprinted in Pareto 1966, p. 163).
92. Ibid., p. 165.
93. Pareto to Naville, 14th January 1900, see (Pareto 1975a, p. 398). With reference to the Swiss case in general, Pareto was "convinced that the socialists would govern better, or less badly, than the radicals" as was evinced by the fact that the Swiss socialists, in the federal elections of 26th October 1902, had, "probably in plain imitation of the German socialists", voted against the protectionist policy of the federal council with its large radical majority (in Pareto's view the socialists had lost the election, winning only 11 seats of the 167 being contested, precisely because of their free-trade policy), Pareto to Maurice Millioud, 5th November 1902, BPS-la.
94. Pareto to Edouard Tallichet, 7th March 1901, BPS-la. He also considered that socialism had advanced in England with the advent of municipal socialism, the progressive inheritance tax and the alignment of the liberals with the Fabians. The English case also showed that "concessions are to no purpose [as] the beloved citizenry is not minimally satisfied with the abolition of the Corn Laws ...; they want to take money from those who have it and each day they take a little more". Pareto emphasised that he was not saying that all this was good or bad but simply limiting himself to describing the situation, Pareto to Papafava, 20th January 1904, see (Pareto 1989, p. 431). As for Germany, if it managed to resist the social

collapse caused by the socialists, Pareto expected it to "inherit the hegemony of Europe", see (Pareto 1902c).

95. See (Pareto 1900a, p. 161).
96. See (Pareto 1902a, pp. 453–454). In Pareto's view, "the humanitarianism of bourgeois socialism has aided socialism in subduing the conservative opposition but in the long run it may also harm it, acting on it exactly as it has on other opponents", see (Pareto 1905a, reprinted in Pareto 1987, p. 455).
97. See (Pareto 1903). What is more, in his view "what the collectivists desire is to take possession of the goods of the bourgeoisie and make use of them". Pareto to Papafava, 20th January 1904, see (Pareto 1989, p. 431).
98. See (Pareto 1900a, p. 163).
99. Pareto to Papafava, 22nd June 1900, see (Pareto 1989, p. 369).
100. Pareto to Giuseppe Prezzolini, 16th January 1904, see (Pareto 1975a, p. 515).
101. Ibid.
102. See (Pareto 1904a, reprinted in Pareto 1987, p. 435).
103. Ibid.
104. Even though it obtained 21.35% of the votes (against 12.97% in the previous elections held on 3rd and 10th June 1900), the Italian socialist party won only 29 of the 508 seats, 4 less than in the previous legislature, due to the effects of majoritarian electoral law.
105. On this political faction see recently Volpe (2015). In any case, commenting on Giolitti's victory, Pareto remarked that while the bourgeoisie continued essentially to "gull the people" after being "placed in power in popular elections", a new elite was emerging which "is opening people's eyes and is moving to the conquest of the state and, more importantly, of the possessions of that same bourgeoisie", see (Pareto 1904b, reprinted in Pareto 1987, pp. 438–439).
106. See (Pareto 1906b).
107. Ibid.
108. Pareto to Carlo Placci, 6th September 1906, see (Pareto 1975a, p. 573).
109. Pareto to Maffeo Pantaleoni, 4th May 1908, see (Pareto 1984a, p. 91).
110. See (Pareto 1917–1919, §2390).
111. Not only on the European mainland, since in England too, "realistically speaking, imperialism and nationalism count among the few forces that are still capable of keeping socialism in check", see (Pareto 1902a, p. 436, note 1).
112. On the nationalist movement in Italy of the time, see recently (Mazzei 2015).
113. See (Pareto 1901b, reprinted in Pareto 1987, p. 367). He thus considered the traditional reactionary factions to be insignificant. More specifi-

cally, he described Gaëtan Combe de Lestrade as being "a bit soft in the head with his apology for autocracy", Pareto to Pantaleoni, 4th December 1899, see (Pareto 1984b, p. 283), referring to Combe de Lestrade (1896), the last work in praise of the Czarist autocracy published by the French aristocrat, who, in the first semester of the academic year 1900–1901, briefly substituted Pantaleoni in Geneva, see (Busino and Stelling-Michaud 1965, pp. 61–65). Although he fully shared Combes' critical views regarding the consequences of universal suffrage, Pareto thought that Combes was committing a grave error by contrasting the two regimes: "The politicians who swear by universal suffrage will be very happy if the choice is presented between Russian autocracy and universal suffrage because on those terms they are sure to prevail", Pareto to Gaëtan Combes de Lestrade, 12th December 1899, BPS-la. An autocracy which provided "an acceptable government … could constitute a solution to the problem of the government", however, Pareto was "very disappointed that it should be necessary to go back to Trajan" to find an example, ibid. Lastly, Pareto pointed out to Combes that he was underestimating the importance of freedom of thought, of expression, of worship and also, more generally, the fact that "for the majority of civilised people in Western Europe, oppression in these spheres causes untold suffering, worse than the pangs of hunger". Therefore, since "we must take people as they are", it still remained to "find a form of government which would be free of the drawbacks of democratic government while procuring the same advantages for the population", ibid.

114. Pareto to Tancredi Galimberti, 18th February 1902, BPS-la.
115. Pareto to Nicola Fornelli, 18th June 1903, BPS-la.
116. See (Pareto 1905b, reprinted in Pareto 1987, p. 410).
117. Pareto to Prezzolini, 17th December 1903, see (Pareto 1975a, pp. 507–508). However, once again in February 1905, Pareto declared that "at this moment in history the class struggle is unquestionably the most significant issue, to the point that in Russia it can even push patriotism into the background", see (Pareto 1905c, p. 447).
118. The conflict of the Boer colonists of the Transvaal and the Orange Free State with the British began on the 11th of October 1899 and ended with the surrender of the Boers on the 31st of May 1902. Pareto declared that he was against this war, but indicated that it had been the Boers' intention "in Africa to play the role of Prussia in Germany or of Piedmont in Italy", Pareto to Moneta, 9th March 1900, see (Pareto 2001, p. 126). It is of interest to note that Pareto's position was contrary to the pro-Boer one adopted by the international pacifist movement and by the bulk of Swiss public opinion, corresponding instead to that of the Anglophile minority who insisted on the desirability of maintaining the traditional excellent

relations between Britain and Switzerland, while denouncing the expansionist and racist character of the Boer political regimes, see (Grossi 1994, pp. 220, 234, 250–252).

119. Pareto to Prezzolini, 17th December 1903, see (Pareto 1975a, pp. 507–508).
120. In this regard see (Mangoni 1981).
121. Pareto to Papafava, 2nd February 1904, see (Pareto 1989, p. 432).
122. A need which "socialism has forgotten since it became transformist and reformist", see (Pareto 1913a, reprinted in Pareto 1987, p. 519).
123. See (Pareto 1913b, reprinted in Pareto 1987, pp. 503–504). Towards the end of the war Pareto was to claim not to be "a nationalist, but a simple onlooker" who had "foreseen that the various schools of thought were about to be engulfed by nationalism", Pareto to Enrico Bignami, 24th July 1918, BPS-la.
124. Pareto to Secrétan, 30th January 1904, BPS-la. Secrétan, in his article *Universal peace (La paix universelle)*, "Gazette de Lausanne", 30th January 1904, had claimed that the spokesmen of the ongoing antimilitaristic campaign in Switzerland, in announcing "a reign of justice and universal peace, where the peoples will all love each other and humans will be brothers", showed that they were living "in a dream world" since "people everywhere continue to fight". He also pointed out that war is not at all incompatible with universal suffrage and that neither did the socialist parties, when in power, hesitate to implement protectionist and nationalist policies.
125. Shortly after this Pareto, on whose antimilitaristic phase see (Mornati 2018a, §5.3), undertook a critical review of some of the prevailing themes of international pacifism, concluding that "the Hague court, if it can ever be of service, will serve to resolve minor questions which would certainly never lead to a war and the same applies to the much-lauded arbitration treaties", see (Pareto 1905d, reprinted in Pareto 1987, p. 457). Later, in reference to the Libyan campaign, he pointed out sarcastically that "these esteemed [Italian] pacifists who had never ceased to acclaim the arbitration process and the court of the Hague, which prescribed going through this same process of arbitration before turning to war, then elected a government which did not care a jot about it", see (Pareto, §1709).
126. Pareto to Placci, 11th February 1904, see (Pareto 1975a, p. 518).
127. See (Pareto 1904c, reprinted in Pareto 1987, pp. 414–415). Similarly, see (Pareto 1904b, reprinted in Pareto 1987, p. 439). For the first sketch of the idea that socialism's advance could be halted only by "long wars between the civilised peoples" see (Pareto 1900b, reprinted in Pareto 1980, p. 238).

128. Pareto to Moneta, 24th September 1904, see (Pareto 2001, p. 160). However, in April 1909, following the recent annexation of Bosnia-Herzegovina by Austria-Hungary in October 1908, Pareto, in keeping with his new focus on the place of force in society, which will be further explored in section 5.7 below, repeated that likewise "in international relations force is the only thing that counts", Pareto to Linaker, 2nd April 1909, see (Pareto 1975a, p. 656), adding he "had been very pleased to see Austria's courage by comparison to the cowardice of the European powers", Pareto to Pantaleoni, 20th March 1909, see (Pareto 1984a, p. 133).
129. See (Pareto 1905e, reprinted in Pareto 1988, p. 45).
130. Pareto to Ernest Roguin, 20th March 1914, see (Pareto 1975b, p. 864).
131. Pareto to Pantaleoni, 14th August 1914, see (Pareto 1984a, p. 173).
132. See (Pareto 1905d, reprinted in Pareto 1987, p. 459).
133. See Pareto 1905f, reprinted in Pareto 1987, p. 465).
134. Pareto to Sorel, 11th November 1909, see (Pareto 1975b, p. 681).
135. Pareto to Teodoro Moneta, 12th October 1911, BPS-la.
136. See (Pareto 1917–1919, §1707).
137. Ibid.

Bibliography

Busino, Giovanni, and Sven Stelling-Michaud. 1965. Materiaux pour une histoire des sciences sociales à Genève. Lettres de Pareto, Pantaleoni, Einaudi, d'Adrien Naville et d'autres [Materials for a history of the social sciences at Geneva. Letters of Pareto, Pantaleoni, Einaudi, d'Adrien Naville and others]. *Cahiers Vilfredo Pareto*, III-6.

Ciuffoletti, Zeffiro, Maurizio Degl'Innocenti, and Giovanni Sabbatucci. 1992. *Storia del PSI. Le origini e l'età giolittiana [History of the Italian socialist party. Its origins and the era of Giolitti]*. Rome-Bari: Laterza.

Combe de Lestrade, Gaëtan. 1896. *La Russie économique et sociale à l'avènement de S.M. Nicolas II [The Russian economy and society at the accession of H.R.H. Nicholas II]*. Paris: Guillaumin.

Croce, Benedetto. 1898. Essai d'interpretation et de critique de quelques concepts du marxisme [Essay of interpretation and criticism of certain Marxist concepts]. *Le devenir social*, February, pp. 97–109.

———. 1899a. Recenti interpretazioni della teoria marxistica del valore e polemiche intorno ad esse [Recent interpretations of Marx's theory of value and the polemics surrounding them]. *La Riforma sociale*, pp. 413–426.

———. 1899b. Una obiezione alla legge marxista della caduta del saggio di profitto [An objection to the Marxist law regarding the decline of rate of profit]. *Atti dell'Accademia Pontaniana*, pp. 199–207.

Grossi, Verdiana. 1994. *Le pacifisme européen 1889–1914 [Pacifism in Europe 1889–1914]*. Brussels: Bruylant.

Mangoni, Luisa. 1981. Le riviste del nazionalismo (The nationalist reviews). In *La cultura italiana tra 800 e 900 e le origini del nazionalismo [Italian culture between 1800 and 1900 and the origins of nationalism]*, 273–302. Florence: Olschki, *passim*.

Mazzei, Federico, ed. 2015. *Nazione e anti-Nazione 1, Il movimento nazionalista da Adua alla guerra di Libia [Nation and anti-nation 1, the nationalist movement from Adua to the war in Libya]*. Rome: Viella.

Mornati, Fiorenzo. 2018a. *An intellectual biography of Vilfredo Pareto, I, from science to liberty (1848–1891)*. London: Palgrave Macmillan.

———. 2018b. *An intellectual biography of Vilfredo Pareto, II, illusions and delusions of liberty (1891–1898)*. London: Palgrave Macmillan.

Pareto, Vilfredo. 1899a. [Review of Arturo Labriola. *La teoria del valore di Carlo Marx: studio sul terzo libro del Capitale [Marx's theory of value. A study of the third book of Das Kapital]*. Milan-Palermo: Sandron]. *Zeitschrift für Sozialwissenschaft*, pp. 546–547.

———. 1899b. [Review of Vincenzo Giuffrida. *Il terzo volume del Capitale di Carlo Marx [The third volume of Karl Marx's Das Kapital]*. Catania: Giannotta]. Zeitschrift für Sozialwissenschaft, pp. 849–941.

———. 1899c. [Review of Georges Renard. *Le régime socialiste. Principes de son organisation politique et économique [The socialist regime. Principles of its political and economic organisation]*. Paris: Alcan]. *Zeitschrift für Sozialwissenschaft*, pp. 148–152.

———. 1899d. La marée socialiste [The socialist tide]. *Le Monde économique*, December 16, pp. 769–771.

———. 1900a. Le péril socialiste [The socialist peril]. *Journal des économistes* LIX (XLIII-2): 161–178.

———. 1900b. Un'applicazione di teorie sociologiche [An application of sociological theories]. *Rivista italiana di sociologia*, July, pp. 401–456.

———. 1901a. *Les systèmes socialistes [Socialist Systems]*, tome I. Paris: Giard et Brière.

———. 1901b. Un poco di fisiologia sociale [A word about social physiology]. *La vita internazionale*, September 5, pp. 529–532.

———. 1902a. *Les systèmes socialistes [Socialist Systems]*, tome II. Paris: Giard et Brière.

———. 1902b. Nouvelles castes sociales (New social castes). *Gazette de Lausanne*, September 25.

———. 1902c. Le tarif douanier allemande [The German customs tariff]. *Gazette de Lausanne*, December 4.

———. 1903. Socialistes transigeants et intransigeants [Moderate and hardline socialists]. *Journal de Genève*, August 17.

———. 1904a. Umanitari e socialisti [Humanitarians and socialists]. *Il Regno*, October 30, pp. 1–2.
———. 1904b. Memento homo [Remember, man]. *Il Regno*, December 11, p. 3.
———. 1904c. Perché? [Why?]. *Il Regno*, February 21, pp. 2–3.
———. 1905a. Socialismo legalitario e socialismo rivoluzionario [Lawful socialism and revolutionary socialism]. *Il divenire sociale*, avril 1, pp. 107–108.
———. 1905b. Il crepuscolo della libertà [The twilight of liberty]. *Rivista d'Italia*, February, pp. 193–205.
———. 1905c. A proposito dei fatti in Russia [On events in Russia]. *Il Regno*, February 12, pp. 5–6.
———. 1905d. Logica umanitaria [Humanitarian logic]. *Il Regno*, June 25, pp. 3–4.
———. 1905e. Guerra [War]. *L'Idea liberale*, September 3.
———. 1905f. Di tutto un poco [A little bit of everything]. *Il Regno*, December 16, pp. 2–3.
———. 1906a. *Manuale d'economia politica con una introduzione alla scienza sociale [Manual of political economy with an introduction to social science]*. Milan: Società Editrice Libraria.
———. 1906b. Le syndicalisme [Trade Unionism]. *Gazette de Lausanne*, September 7.
———. 1913a. Sul nazionalismo [About the nationalism]. In *Il nazionalismo giudicato da letterati, artisti, scienziati, uomini politici e giornalisti [Nationalism in the judgment of literati, artists, scientists, politicians and journalists]* with a preface by Arturo Salucci, 187–188. Genoa: Libreria Editrice Moderna.
———. 1913b. 'Referendum' sulla massoneria ['Referendum' on freemasonry]. *L'Idea nazionale*, September 18.
———. 1917–1919. *Traité de sociologie générale [Treatise on general sociology]*. Lausanne-Paris: Payot.
———. 1966. *Mythes et Idéologies [Myths and ideologies]*. Complete Works, tome VI, ed. Giovanni Busino. Geneva: Droz.
———. 1975a. *Epistolario 1890–1923 [Correspondence, 1890–1923]*. Complete works, tome XIX-1, ed. Giovanni Busino. Geneva: Droz.
———. 1975b. *Epistolario 1890–1923 [Correspondence, 1890–1923]*. Complete works, tome XIX-2, ed. Giovanni Busino. Geneva: Droz.
———. 1980. *Écrits sociologiques mineurs [Minor sociological writings]*. Complete works, tome XXII, ed. Giovanni Busino. Geneva: Droz.
———. 1984a. *Lettere a Maffeo Pantaleoni 1907–1923 [Letters to Maffeo Pantaleoni 1907–1923]*. Complete works, tome XXVIII.III, ed. Gabriele De Rosa. Geneva: Droz.
———. 1984b. *Lettere a Maffeo Pantaleoni 1897–1906 [Letters to Maffeo Pantaleoni 1897–1906]*. Complete works, tome XXVIII.II, ed. Gabriele De Rosa. Geneva: Droz.

———. 1987. *Écrits politiques. Reazione, Libertà, Fascismo, 1896–1923 [Political writings. Reaction, Liberty, Fascism, 1896–1923]*. Complete works, tome XVIII, ed. Giovanni Busino. Geneva: Droz.

———. 1988. *Pages retrouvées [Rediscovered pages]*. Complete Works, tome XXIX, ed. Giovanni Busino. Geneva: Droz.

———. 1989. *Lettres et Correspondances [Letters and correspondence]*. Complete works, tome XXX, ed. Giovanni Busino. Geneva: Droz.

———. 2001. *Nouvelles Lettres 1870–1923 [New Letters 1870–1923]*. Complete Works, tome XXXI, ed. Fiorenzo Mornati. Geneva: Droz.

Volpe, Giorgio. 2015. *La disillusione socialista. Storia del sindacalismo rivoluzionario in Italia [The socialist disillusion. History of revolutionary unionism in Italy]*. Rome: Edizioni di storia e letteratura.

CHAPTER 5

A New Sociology

As of 1907 Pareto began to voice his conviction that "economics is simply a branch of sociology",[1] with the suggestion that "purely economic deductions are fairly remote from reality"[2] and therefore provide "only one, often secondary, factor for the resolution of issues which, this notwithstanding, are nevertheless termed economic".[3] Then, in 1913, Pareto confessed that where in the past he had thought "that economics could be studied independently of sociology",[4] he now believed that "it is essential to connect economic phenomena with other social phenomena in order to arrive at a theory covering situations arising in the real world".[5]

Consequently, it amounted simply to further confirmation of this when, on the 6th of July 1917, for the occasion of the official ceremony marking the 25th anniversary of his teaching activities in Lausanne, he recalled that "having arrived at a certain point in [his] research in political economy", he found himself at an "impasse", prevented from "gaining access to empirical" reality by the fact that "the mutual interdependence characterising social phenomena prevents their various ramifications from being examined in isolation".[6] Accordingly, his *Treatise on General Sociology*[7] was written as an "essential complement for the study of political economy".[8] Thence, in this chapter we will retrace what appear to constitute the essential steps towards the original conception of sociology which Pareto succeeded painstakingly in formulating, comprising the theory of action (§1), social heterogeneity (§2), social equilibrium (§3) and some elements of a

F. Mornati, *Vilfredo Pareto: An Intellectual Biography Volume III*,
Palgrave Studies in the History of Economic Thought,
https://doi.org/10.1007/978-3-030-57757-5_5

sociology of politics (§4). This will be followed by an appendix dedicated to the multifarious publishing projects connected to the *Treatise*.

5.1 The Structure of Action

It was at the end of 1898 that Pareto first contrasted actions performed "intentionally" to those performed "out of habit", insisting that people tend to attribute to the former category actions which in fact belong to the second, while at the same time seeking to assign "more or less convincing motives" to them.[9] In July 1900, Pareto then reached the further, and for him equally fundamental, conclusion that "the majority of human actions originate not in reason but in feelings",[10] that is, they are, according to the terms of the distinction made above, habitual, "a fact which [for that matter] has been recognised for centuries".[11]

In his *Treatise on Sociology*, Pareto returned to his original distinction and refined it. With the proviso[12] that from the point of view of the individual concerned "almost all actions are logically connected to the goal",[13] Pareto characterised as logical those actions which are logically connected to their goal, not only in the mind of the person involved "but also for those with a broader perspective".[14] By contrast, the remaining actions are non-logical "which does not mean illogical".[15] In other words, the objective purpose (meaning "within the realm of observation and experience"[16]) of logical actions coincides with their subjective aim,[17] while non-logical actions[18] are those whose objective purpose does not coincide with their subjective aim or which have neither an objective nor a subjective purpose, or which have an objective (subjective) purpose but not a subjective (objective) one.

Pareto then specified that logical actions "are, at least predominantly, the result of a reasoning process", while non-logical ones "derive principally from a certain psychological state", the origins of which Pareto left to the domain of psychology.[19]

This psychological state,[20] in turn, not only engenders the action but also the theory that the person performing the action believes the true origin of the action.[21] On the other hand, the theory and the action can reinforce the psychological state which engendered them.[22] As a biographically significant example of this mechanism,[23] Pareto cites a certain combination of "economic, political and social interests of the individuals and the circumstances in which they live" which engenders a psychological state which, in turn, separately begets both the application and the theory of free trade, where the theory's influence on the practical application of free trade is "very weak".[24]

It is possible to ascertain a psychological state from the theory and the action which it engenders.[25] For the purposes of this study, Pareto saw the theories as being characterised both by the empirical or non-empirical nature of their constituents and by the logical or non-logical nature of the links between these constituents. Consequently, there are empirical and logical theories, empirical and non-logical theories, non-empirical and logical theories, and non-empirical and non-logical theories.[26] Pareto's investigation of these theories was conducted in the light of the fact that "every scholar who has reflected and meditated on social phenomena has concluded that there is a variable component and also another component which is relatively constant".[27] Hence, through an analysis of the theories, Pareto sought to identify the constituents of non-logical actions,[28] constituents to be classified "into orders, classes, genera and species", just as in botany.[29]

Following a lengthy review of examples taken from Greek and Roman mythology,[30] Pareto concluded that in logical/empirical theories, it is possible to identify "over and above the factual data, an essential component and a contingent component which is generally quite variable". The former "is the principle which exists in the human spirit", while the latter consists of the "explanations [and] deductions from this principle".[31] In other words, the first represents "the expression of certain feelings [and] corresponds directly to non-logical actions",[32] while the second "is the manifestation of man's need for logic"[33] and corresponds to derivations. In Pareto's view, the former "is more important in the search for social equilibrium"[34] even if the latter, "while secondary, also has a bearing on equilibrium".[35]

It was solely for purposes of explanatory convenience that Pareto termed the principal parts of theories residues, the secondary parts derivations[36] and the combinations of residues and derivations derivates,[37] specifying that "residues are the manifestations of the [corresponding] feelings and instincts".[38]

On this basis, Pareto concluded that there are six categories of residues: the instinct for combinations (divided into six types), the persistence of aggregates (seven types), the need to express one's feelings through external actions[39] (two types), residues associated with sociability (six types[40]), the integrity of the individual and of his dependencies (four types[41]), and the sexual residue (no sub-types).[42]

As to the social importance of residues, Pareto had already noted in the sociological section of the *Manual of Political Economy* that "people who live in society possess certain feelings which, under specific circumstances, guide their actions". These can be identified more specifically as religion, morality, law and custom.[43] Pareto thus considered that it would be extremely important "to know how feelings arise, develop and recede" but, on the contrary, little is known about it.[44] In any case, feelings appear to change by imitation between individuals, social classes and peoples.[45] Imitation, however, is contrasted by opposition, in accordance with the general rule whereby "when a doctrine is generally accepted, an adversary will materialise to oppose it". Thus, in that period, humanitarianism and egalitarianism were counterbalanced by Nietzsche's notion of the egotistical superman.[46]

Of the residues, religious sentiment (belonging to the category of persistence of aggregates) is the one to which Pareto devoted most attention in the period prior to the *Treatise*. Indeed, since 1900 he had observed that the contemporary importance of religious sentiment demonstrated "its significant role in the maintenance of social order", but that it was not possible to specify how much or what type of religious sentiment led to the maximum of social utility.[47] In 1908, nevertheless, he had reached the conclusion that "in social terms the best type of religion continues to be that involving the supernatural".[48]

The analysis of residues needs also to take into account the degree and the frequency with which they manifest themselves in society,[49] as well as the way they change over time, between social classes and across society as a whole.[50] Thus, for example, in Pareto's opinion, the role of the instinct for combinations in human actions has changed little over time[51]; however, for the category of the persistence of aggregates,[52] in Italy, between 1908 and 1911, there had been a "very remarkable transformation from pacifist to warlike religious belief".[53] In relation to the category of the need to manifest feelings through external actions, among the civilised peoples of the time Christian cults had been substituted, at least in part, by the cult of the State and of the People,[54] while, contrary to popular belief, in Pareto's opinion the residues relating to socialising (i.e. sociability) seemed to have diminished by comparison to the residues for the integrity of the individual and of his dependencies (i.e. individualism), as demonstrated by the increased power of the Trade Union bosses and the tolerance shown towards lawbreakers[55]; finally, "sexual residues show possibly the least variation".[56]

Pareto also considered that "the variability [of residues and derivations] is more apparent in the various social strata than in the society as a whole, where compensations operate between the different strata",[57] that "residues have a powerful action on derivations and derivations a weak action on residues",[58] that conflicting residues have corresponding derivations which are similarly conflicting[59] and that a residue interacts more easily with other residues of the same type than with residues of different types.[60]

5.2 Social Heterogeneity

If we take as valid the top-shaped pattern of wealth distribution proposed by Otto Ammon, which Pareto thought he had shown to vary little in time or space, it "represents the outward configuration of the social organism".[61] The positioning of people "according to their degree of influence and power", in fact, not only reproduces the pattern of income distribution but normally shows the same few individuals at the apex.[62]

Pareto then remarked that human society is not homogeneous, in the sense that it is composed of individuals who differ in characteristics which are visible[63] to a greater or lesser degree. To be more precise, each quality modulates "imperceptibly" from the many who possess it to a modest extent to the few who possess it to a considerable extent.[64]

Consequently, since human beings "differ significantly from one another physically, morally and intellectually", in Pareto's view "human society can be characterised as a collective with a hierarchy",[65] even if one should always keep in mind that élites do not endure[66] but are constantly renewed.[67]

More specifically, Pareto observed that a society "of homogeneous nationality" may be divided into two groups A and B, which are "openly hostile" to each other with the first dominating[68] the second, while a third group C can also be introduced which will sometimes support A and sometimes B.[69] Further, group A is internally divided into a subgroup Aα, which is able to maintain itself in power and a subgroup Aβ, which, conversely, is composed of degenerate individuals; group B, on the other hand, is subdivided into a group Bα, which "constitutes the emerging new aristocracy",[70] and a group Bβ, which is made up of the "vulgar mob which forms the bulk of human society".[71]

On this basis, Pareto observed that "objectively", the political contest is between the Aα and the Bα, with both groups looking for the supporters needed to gain victory[72] and being able to gain recruits only by disguising the fact that they are interested solely in power. Hence, the Aα group seek to appear as the guarantors of the common good, while the Bα group attempt to present themselves as the defenders of Bβ.[73] Consequently, the political battle, which is simply a power struggle, takes on the appearance of a confrontation over the great principles of liberty, law and equality.[74]

In so doing, the Aα alienate the Aβ but procure only the temporary backing of the Bβ,[75] while the Bα win the definitive endorsement not only of the Bβ but also of some of the C and of the majority of the Aβ,[76] who, despite being "degenerate", are nevertheless of a higher calibre than the Bβ and are supplied with money, always an indispensable element in the financing of conflicts.[77]

Thus, the Aβ are the initiators of all revolutions,[78] which culminate, however, in the victory of the Bα, who overcome the Aα by force, in the disappointment of the Aβ group, who never achieve their aims and, lastly, in the procurement on the part of the Bβ of some minor rewards both during the conflict and then with the change of overmaster.[79]

Pareto also developed two other representations of the social hierarchy. If one could assign "to each individual in every branch of human activity an index showing his capabilities",[80] then the term "elite" would refer to the group of individuals having the highest index in the sector where they are active.[81] For the purposes of the study of social equilibrium, Pareto thought it expedient for the elite to be divided into two groups, with the governing elite comprising those in the government or directly influencing its decisions and the non-governing elite comprising the rest.[82] Pareto added that it was also possible to join the elite, particularly the governing elite, on the basis of "wealth, family or connections".[83]

Pareto also divided society into "an upper echelon, to which those in government normally belong, and a lower echelon comprising those being governed".[84] Pointing out that the governing class benefits from the acquisition of new members arriving from the lower class who bring with them "the energy and the quantities of residues needed to maintain itself in power" and also from the loss of its "most degenerate" elements,[85] Pareto underlined that the presence of either of these will destroy the governing class and the whole country with it. This is because the accumulation of superior individuals among the governed or of unsuitable individuals among the governing class will lead to the upsetting of the social equilibrium in revolutions, instigated, necessarily with accompanying use of force, by those wishing to remove from power anyone lacking the capacity to maintain it.[86]

Finally, Pareto also concluded in relation to social heterogeneity that to obtain the maximum social utility, the different social classes should follow different codes of behaviour. However, due to the strong bonds of fellowship and blending between them, this could be achieved only by adopting a principle common to all and then by making "all the necessary exceptions, thanks to which the principle would be general only in appearance".[87]

5.3 Social Equilibrium

From the outset, Pareto's deliberations on sociology included a variety of endeavours to put the various notions he had developed into juxtaposition.

As early as March 1899, he remarked, in reference to the relations between the political and economic systems, that "a close connection" had been shown to exist between them, "[which] is completely different" from showing that the former is a "consequence" of the latter.[88]

In the *Manuel d'économie politique (Manual of Political Economy)* Pareto stated that among the various factors "contributing" to the humanitarian reorientation of social sentiments which had been seen in France around the turn of the century, there had been the growth in national income (which made it possible to devote a part to various forms of humanitarianism), together with the growing involvement of the poorer reaches of society in government, the decadence of the bourgeoisie and, lastly, 34 years of peace.[89] Pareto also claimed that "the hierarchy, turnover among the aristocracy, social selection and the average amount of wealth or capital per head" to be the factors determining the other elements of the social context within what was, nonetheless, a relationship of mutual dependency.[90]

In conclusion, Pareto's conception of the social system emerges clearly in the *Treatise*. With residues (a), interests (b),[91] derivations (c), social heterogeneity and turnover (d) being the elements[92] whose interactions determine "the concrete state of equilibrium" in society,[93] Pareto reiterated that since this pattern could not be expressed in a system of equations because it was still not possible to "assign specific indices to each of mutually-dependent elements",[94] it could be investigated solely through verbal descriptions of "the actions and reactions among these factors which succeed one another in a cycle, ad infinitum".[95]

The following sequences are thus observed[96]:

Sequence I, wherein (a) acts on (b), (c) and (d), "determines a very extensive component of the social scenario".

Sequence II, wherein (b) acts on (a), (c) and (d), "determines a significant component of the social scenario". In particular, this was understood by the exponents of historical materialism "who, unfortunately, fell into the error of neglecting the other combinations".

Sequence III, wherein (c) acts on (a), (b) and (d), "is of lesser importance than all the others".

Sequence IV, wherein (d) acts on (a), (b) and (c), "is not without importance", as shown by the fact that ignoring it "radically invalidates the so-called democratic theories".

Thus, for example, a variation in (a) for sequence I causes variations in (b), (c) and (d) which in turn gives rise, via sequences II, III and IV, to a further variation in (a), and so on.[97]

An initial example of extensive sociological analysis is represented by customs protection,[98] that is, a variation in (b) in sequence II which, on the basis of its direct economic effects alone, leads to a destruction of wealth. Among non-economic effects of a variation in (b) in sequence II, "the most important applies to (d)", in the sense that protection enriches and increases the political power of the industrialists benefitting from the protection and of the politicians delivering it to them, in all of whom the instinct for combinations prevails over the persistence of aggregates.[99] On the other hand, the effects of protection on residues are negligible because these "are slow to change", while "significant effects are felt by (c), typically in the guise of a remarkable proliferation of economic theories in defence of protection".[100]

Turning to (c), a variation here will, in sequence III, have no effect on the residues, a limited effect on interests but a more significant one on (d), "because in every society, people who are able to suck up to the powerful will be able to obtain a place in the governing class".[101]

Lastly, these variations in (d) will lead, in sequence IV, to negligible variations in the residues but very considerable effects on the interests.[102] It can indeed happen that customs protection brings into the government "people well provided with class I residues who spur the whole nation towards industrialism",[103] thus leading, particularly when the turnover of the élites suddenly accelerates sharply after a period of stagnancy,[104] to an increase in production which can more than compensate the destruction of wealth caused initially by protectionism.[105] It can therefore arise, in contradiction of everything that Pareto had long expounded on the question, that "a country's economic prosperity increases with industrial protection [too]".[106] In January 1920, after declaring that "the greatest obstacle to the functioning of Italian industry is the impositions of the workforce", observed that customs protection, despite leading in material terms to the destruction of wealth, "is to be rejected" only if "it brings no benefit to the nation", which is the case for the "production of war materials ... in its principal form [i.e. the construction of ships] for the navy".[107] After the war, he accepted that customs protection introduced without any evident "benefit to the nation serves simply to swallow up profits".[108] Again in January 1921, Pareto conceded that the liberal accusation against protectionism of wishing to produce everything in one country "is no longer so clear-cut for those nations possessing a variety of climate and range of output which allow them to isolate themselves from other nations while suffering no major penalty" and indeed "loses all its force if the political side of the question is taken into account", whereby "isolation can constitute the only means of escaping from foreign domination".[109]

A further example of a broad sociological analysis starts out from the notion that the instinct for combinations predominates in entrepreneurs while the persistence of aggregates predominates in savers,[110] with these two classes having opposing economic interests (e.g. the entrepreneurs seek to pay as little as possible for the savings they take in the form of loans while the savers seek to gain as much as possible).[111] More generally, while entrepreneurs belong to the category of speculators, consisting of those "whose income is essentially variable and depends on their skill in identifying sources of profit",[112] savers are in the category of beneficiaries of annuities (or rentiers), that is, those "whose revenue is fixed and depends little on the ingenious schemes that they can dream up".[113] Again, from the social point of view, the category of speculators "is the principal engine of change and of social and economic progress", whereas the category of rentiers "is, on the contrary, a powerful force for stability". As a result, a society dominated by speculators "lacks stability", while a society dominated by rentiers "remains stationary".[114]

From the perspective of social dynamics, a country's economic, political and social prosperity appeared to Pareto to resemble an "inverted U" depending on the ratio between the instinct for combinations and the persistence of aggregates and implying that this prosperity has a maximum value when the ratio is intermediate between its minimum and maximum values (scant instinct for combinations and ample persistence of aggregates and ample instinct for combinations and scant persistence of aggregates, respectively).[115] Pareto thought that the societies of his time, after a period of prosperity, went into a period of decline owing to an excess of the instinct for combinations.[116]

Nevertheless, as in the past, a period of renewed prosperity could still be anticipated for the societies of the time, were a revolution to put power in the hands of individuals in abundant possession of residues of the persistence of aggregates who, additionally, "had the ability, the opportunity and the desire to make use of force".[117]

Pareto added that periods of prosperity are favourable to speculators, whose increased wealth facilitates their entry into the governing class, and unfavourable to rentiers, partly because inflation reduces their real income and partly because they are unable to counter the speculators' success with the governing class and with the public. Vice versa, periods of "economic stagnation" are favourable for the rentiers and unfavourable for the speculators.[118] Hence, during periods of prevailing prosperity, the residues of the instinct for combinations, characteristic of speculators, will be reinforced while the residues of the persistence of aggregates, typifying rentiers, will decline, "which has the effect of progressively spurring the various peoples to engage in economic enterprise and in the advancement of economic prosperity, until the time where new forces emerge which neutralise the process". During periods of economic stagnation this mechanism is reversed.[119]

A third case of wide-ranging sociological analysis concerned the issue of social stability.

Already back in 1901 Pareto had asserted that a society can survive only, firstly, if mechanisms are in operation to "counteract factors which contribute to its decay or even simply to its debilitation"[120] and, secondly, if some form of "mutual assistance" exists among its citizens.[121] He later explored the question further, adding that in modern societies the necessary element of stability was provided by private property and inheritance, while the equally fundamental force for change emerged from "the right of each person to rise as far as possible in the social hierarchy". Pareto added further that "it is impossible to determine a priori whether it would be beneficial or damaging to society" to modify its condition of stability and/or of volatility.[122] In the end he concluded that "the main function of feelings of the persistence of aggregates is to counteract the deleterious effects of selfish interests and unrestrained passions in an effective manner". More specifically, the persistence of aggregates is of use to society when it is felt strongly, particularly "in subordinates",[123] becoming harmful when it loses its power.[124]

5.4 Elements of the Sociology of Politics

According to Pareto it was obvious that the members of a society have only certain interests in common,[125] meaning that there is always a conflict between the interests of a minority and those of a majority.[126]

The existence of such a conflict of interests leads to political action[127] which, according to Pareto, should set achievable goals and then should pursue them in the most effective manner. Thus, those wishing to preserve society (bearing in mind that "very often governments fall not because of the strength of their opponents but rather because they lack the courage to defend themselves"[128]) need to stand by any strong government and fight against any ideology which threatens it, while vice versa, "those wishing to pave the way for revolution" should stand by any weak government and fight against any strong government and its supporting ideologies.

In Pareto's view, the first task of any government is its own preservation, which is pursued by making use, firstly, "of appropriate methods" in the context of the prevailing situation, methods which comprise "an inevitable mixture of good and bad"[129] and, subsequently, by identifying a "way" to grant privileges at the expense of the rest of society,[130] privileges introduced through legislation which will be sweeping and partisan in proportion with the extent of state power. Consequently, the workings of a government, as for any other type of social organism, can be judged only through an ongoing process of assessment, in itself inevitably highly approximate, together with the subsequent evaluation of whether positive or negative factors predominate at each stage.[131]

In general, those in government direct a much larger group of citizens, "partly through force and partly by consent",[132] in varying proportions[133] and with varying methods of eliciting consent or of applying force.[134]

Specifying that while cases of unanimous consensus have never yet been recorded, with the opposite case of a regime which maintains itself through force alone being more common,[135] Pareto observed, as regards methods of eliciting consent, that governments are more effective when, like

Giolitti,[136] they seek to make use of existing residues than when, as in the case of Crispi,[137] they try to suppress these, sometimes in a violent manner.[138] It is also possible to elicit consent by giving priority to interests, bearing in mind, however, that this kind of protection, if not combined with fervour, can serve for the reconciliation of those in whom the instinct of combinations prevails (and thus who largely already belong to the governing class) but not of those in whom the persistence of aggregates is dominant (i.e. those belonging to the much larger category of the governed).[139] The use of force, itself, is in the hands of the armed forces with the further patronage of the political clienteles[140] and the former can be more or less costly than the latter.[141]

On this basis Pareto recognised the existence of two types of government. The first corresponds to those governments which rely principally on force together with religious sentiment. These are less costly and do not stimulate economic production, since they are led by élites with a low turnover, in whom the residues of the persistence of aggregates predominate.[142] According to Pareto, these governments, when they are provided with a certain quantity of the instinct for combinations, can be long-lasting.[143] The second type corresponds to governments which seek to promote interests, making use of various forms of trickery. These can be very costly, but they greatly stimulate economic production because they are led by high-turnover élites in whom residues of the instinct for combinations predominate.[144]

Pareto then observed that within governing parties, two basic tendencies exist.[145] One is characterised by a prevalence of residues of the persistence of aggregates and is thus made up of individuals who "resolutely target idealistic goals". The second shows a prevalence of residues of the instinct for combinations and is consequently made up of individuals "whose goal is to seek their own interests and those of their clients". In Pareto's view, the fact of possessing an abundance of instincts for combinations renders this group "[the] best suited to govern" but added further "it is no accident when power is given to a dishonest man but is rather a choice prompted by the system".

On the other hand, having established that whatever the form of government, those in power also displayed a tendency "to abuse it in order to gain advantages and particular benefits"[146]; again, in Pareto's[147] view, the abuses by those in power increase in accordance with the level of abuses among the people to which they belong and with the level of state interference in private business.

Further to the distinction made above between rentiers and speculators[148] in society, Pareto also claimed that the former were conservatives or nationalists who lent "stability to nations", who were against inflation (as reducing their standard of living) and against increases in taxation (being unable to pass the costs on), while the latter were innovators and internationalists, bringing "progress" to their nations. These were in favour both of inflation (allowing them to increase prices, which they largely controlled, more than proportionally to the rate of inflation) and of tax increases (because they were able to pass these on, at the same time benefitting from the increases in public expenditure financed by the increase in tax revenues).[149]

However, the category of speculators, despite controlling the entire civilised world of the time,[150] "may know how to scheme, but they do not know how to fight".[151] Their current interests included benefitting from the highest possible level of expenditure on arms,[152] even if they also feared a war because "a victorious general"[153] could modify the balance of political power or because their country might be defeated by an enemy nation where the ratio of rentiers to speculators more closely approximated the optimum for maximising military power.[154]

On the other hand, Pareto emphasised that the role of the rentiers in society was highly important because "civilisation is a direct result of the quantity of savings a people possesses or can bring into play".[155] However, they too, being "generally reserved, prudent and timid people who shun risky undertakings",[156] shied away from the use of force so that a third group, comprising individuals who did not hesitate to make use of force, could prey on the rentiers even more easily than the speculators could.[157]

After the war, reiterating that the social order oscillates between consensus (obtained through shared interests acting with the residue of the persistence of aggregates[158]) and the use of force, Pareto recalled that with their participation in government, the landowners promote the persistence of aggregates while the speculators promote the instinct for combinations.[159] A society characterised by the persistence of aggregates "can maintain itself by its own virtues",[160] whereas a society based on the instinct for combinations, having "little strength of its own", must be sustained by other forces, taking the forms of a demagogic or military plutocracy, with "the first being economically more costly than the second except where the latter exaggerates with its military ventures".[161] Both of these forms of society contain "within themselves the germs, first of prosperity and later of decadence"[162] and in the past "great social upheavals brought them to an end".[163]

Appendix: The Publishing Projects Associated with the *Treatise*

It is of interest to re-examine the complex and ambitious publishing projects which Pareto envisaged and which culminated in the *Treatise*.[164]

In July 1904,[165] Pareto informed Pasquale Jannaccone, the editor of the fifth series of the *Biblioteca degli Economisti*, that he was in a position to submit a *Treatise on Political Economy* for inclusion in the fifth series of the Biblioteca, although "making no promises". Pareto[166] added that he had in mind a work "which would only partially duplicate the *Cours* and would be largely new", maintaining that for this possible work "there is no reason to be concerned that it might compete with the *Manual of political economy* [because] this would have the character and the format of a booklet, with a completely different content".

Nonetheless, in November of that year Pareto announced to the editor Rouge that he wished to completely rewrite the *Cours* in five separate "small volumes",[167] one containing a *Précis of sociology (Précis de sociologie)*,[168] one of *Principles of pure economics (Principes d'économie pure)* ("without mathematical formulae", which "will harm its sales"[169]), one of *Mathematical economics (Économie mathematique)* ("which is the basis of the rest of the work"[170]) and one or two of *Applied economics* (*Économie appliquée*).[171] Then, on the 15th of February 1905, Pareto declared to Rouge that he was "now in agreement on all points", assuring him of his intention "to start drafting the first volume".[172]

In April 1905, Pareto clarified his intentions, explaining to Sensini that he had promised a *Treatise on Political Economy* to the Biblioteca (of this project no more was heard) and that he wished to rewrite the *Cours* in the manner agreed with Rouge, with the possible assistance of Sensini himself.[173]

Once again on the 30th of November 1906, Pareto informed Rouge that he was fully engaged with the preparation of the volumes on sociology, on pure economics and on mathematical economics, while the remaining two volumes on applied economics "could be ready well before". In any case he left to Rouge sole competence as regards the advisability in bringing their publication forward.[174]

A few days later he assured Rouge that he would busy himself "actively, to arrange all the details of the work required" for the preparation of the two volumes on applied economics[175] the completion of which he estimated would in any case take "around 18 months".[176] To gain time, Pareto asked Sensini for his help, offering him, in exchange for the 500 francs per volume which Pareto would receive from Rouge, the task of revising the applied part of the *Cours*, consisting of the updating and completion of the statistical data, the review of the factual information in the light of later events, the addition of information regarding subsequent phenomena such as "English municipal socialism, Italian and German municipal socialism and Trusts in the United States". In essence, the two volumes, of which Sensini would appear as co-author, would be obtained by the removal of "all of the mathematical annotations [and] all the sociological part" of the *Cours* and the addition of "sundry items of economic history, not only as regards the past but also the present".[177] A few days later, Pareto told Sensini that in order to avoid possible linguistic difficulties, he might limit himself to including information concerning the case of Italy.[178] After informing Rouge in June 1907 that the work for the preparation of the new edition of the *Cours* "is far from complete",[179] only a few months later he repeated that his manual of sociology,[180] "augmented and, more importantly, corrected, will become the first volume of the new *Cours*"[181] and that he had not yet abandoned the idea of a second edition of the *Cours* in five volumes.[182] However, in February 1912, he finally informed Rouge that he did not hold out "great hopes of ever being able to complete a new edition [of] the *Cours* because the state of [his] health leaves much to be desired".[183]

In the meantime, Pareto had initiated a new publishing undertaking which was more modest but which would turn out to be decisive. On the 1st of April 1906, in response to a request from the Florentine publisher Piero Barbera, Pareto informed him that he did not think "it would be possible to obtain a new Manual of sociology by translating the first volume of the new edition of the *Cours*".[184] However, on the 18th of November 1906, Pareto wrote to Barbera that "the work on the French-language treatise on sociology, of which the Italian manual would be a compendium, is fairly advanced"[185] and that therefore they could begin discussing terms for the publication of this latter work.[186] On the 9th of January 1907, Pareto signed a contract with Barbera whereby he agreed to the publication by the Florentine publisher of a manual of sociology which "could be the condensation of a *Treatise* written in French [and which] will be published before the *Treatise* itself".[187]

In June 1907, Pareto announced that he had begun the Sociology [book] for Barbera, which he conceived as "an entirely new work, which will follow hitherto little-trodden or untrodden paths",[188] adding that he had in mind "to start with the Italian version and then revise it and make a better one in French".[189] At the end of 1910, Pareto told Barbera that his Sociology "is a completely new work" and that he had "realised that in order to write the *Manual* he need[ed] first to complete the *Treatise* for publication in French and the *Manual* will be a compendium of this". The *Treatise* required a lot of research and was at the halfway point in terms of composition. Once this work was finished, "[he would] write the *Manual* in no time".[190]

In July 1912, he informed Barbera that he had completed the *Treatise*, proposing to publish it and subsequently to delay or cancel the publication of the *Manual*.[191] The publisher accepted the first[192] of Pareto's proposals, publishing the *Treatise* in 1916[193] and the *Manual*, which was entitled *Compendio di sociologia generale (Compendium of general sociology)* in 1920. As for the French edition of the *Treatise*, its publication was initially entrusted to Rouge,[194] who pulled out in February 1913[195] to be replaced by the Lausanne publisher Payot.[196] Similarly, the translation, initially proposed to Millioud, was passed after his refusal to Boven by Pareto.[197] The resulting translation contains admittedly sporadic updates to the Italian text up to October 1918.[198]

Notes

1. Pareto to Pantaleoni, 2nd April 1907, see (Pareto 1984a, pp. 26–27).
2. Pareto to Pantaleoni, 26th October 1907, ibid., p. 68.
3. Pareto to Alfonso de Pietri Tonelli, 31st July 1912, see (Pareto 1975a, p. 778).
4. Pareto to Emanuele Sella, 11th June 1913, ibid., p. 832.
5. Ibid., p. 833.
6. See (Pareto 1920a, reprinted in Pareto 1975b, p. 66).
7. As is known, the Italian edition was published by Barbera in 1916 while the French translation (completed under Pareto's close and satisfied supervision by the young jurist Pierre Boven [1886–1968] from Lausanne, who had just obtained his doctorate under Pareto with a thesis entitled *Mathematical applications in political economy (Les applications mathématiques à l'économie politique)*, Lausanne: Rouge 1912), was pub-

lished under the title *Traité de sociologie générale (Treatise on General Sociology)* by the Lausanne publishing house Payot in 1917 (first volume) and in 1919 (second volume). Here we will follow this version, as being the latest to obtain the author's full approval. Among translations in other languages, the best known is *The Mind and Society*, translated by Andrew Bongiorno and published in the United States in 1935. Also recommended is the major critical edition of the *Treatise* published in 1988 in four volumes by Giovanni Busino for the publisher Utet at Turin. See also (Busino 2008), which provides an essential conceptual and bibliographical introduction to Pareto's sociological studies.

8. See (Pareto 1920a, reprinted in Pareto 1975b, p. 69).
9. See (Pareto 1898, reprinted in Pareto 1987a, pp. 103–104). On Pareto's earlier sociological thinking, see (Mornati 2018a, §6.10) and Mornati (2018b, chapter IX).
10. See (Pareto 1900, reprinted in Pareto 1980, p. 179). It is possibly due to this growing conviction that Pareto, ibid., decided to adopt the methodological approach "passing from facts to ideas and [only] subsequently from ideas to facts" which had been proposed by the French physiologist Claude Bernard (1813–1878), see (Bernard 1878).
11. See (Pareto 1900, reprinted in Pareto 1980, p. 198).
12. Since this definitive taxonomy had already been foreshadowed, see (Pareto 1909, chapter II, §3) and (Pareto 1910, reprinted in Pareto 1980, p346).
13. See (Pareto 1917–1919, §150).
14. Ibid.
15. Ibid.
16. Ibid., §151.
17. Ibid.
18. Ibid.
19. Ibid., §161.
20. Which in reality is only that of the *élite*, ibid., §246.
21. Ibid. §162.
22. Ibid. §165.
23. Ibid. §167.
24. Ibid.
25. Ibid. §169. Before beginning this inductive study of the psychological state underlying non-logical actions, Pareto underlined its innovative character, resulting from the manner in which it conflicts, firstly with the tendency to make non-logical actions appear logical and secondly, with the tendency to assign "greater importance to logical actions". Pareto ascribes the first of these tendencies to the prevailing prescriptive bias in sociological studies, ibid., §253, as well as to the fact that it is "much

easier" to give a theoretical account of logical actions than of non-logical ones, because in the first case the theoretician "possesses within himself the instrument which makes logical deductions", while in the second case he must "seek outside himself for the elements" which are required, particularly through "the scrutiny of a broad array of facts", ibid., §262. Pareto thought that this second tendency depended instead on the idea that non-logical actions are "a discreditable phenomenon which should not occur in a well-regulated society", ibid., §265. According to Pareto, ibid., §1768–1769, society exists notwithstanding the fact that the vast majority of actions are non-logical, because "a residue departing from logic can be corrected by a derivation departing from logic, so that the conclusion approximates what is observed".

26. Pareto, ibid., §13. Moreover, also of interest in relation to a theory are the "reasons" why a given individual expresses or accepts it and whether the theory is productive or detrimental for him, ibid.
27. Ibid. §1719.
28. Ibid. §148.
29. Ibid. §116.
30. These investigations fill chapters IV and V of the *Treatise*. In general, Pareto considered that "historical studies complete studies of the present, radically extend[ing] the sphere of the combinations which we are able to study", see (Pareto 1903, reprinted in Pareto 1980, p. 243.
31. See (Pareto 1917–1919, §798).
32. Ibid. §803 Pareto stated, on the other hand, that in logical/empirical theories the essential component consists of "empirical principles, descriptions, empirical assertions".
33. Ibid. §798.
34. Ibid. §800.
35. Ibid. §801.
36. Derivations, described by Pareto in chapters IX and X of the *Treatise*, are divided into four different classes: affirmation (subdivided into three kinds), authority (three kinds), accord with feelings or principles (six kinds), verbal demonstration (five kinds), ibid., §1419.
37. Ibid. §868.
38. Ibid. §875.
39. In this category of residue "the need to act constitutes the main factor" while feelings "are secondary", ibid., §1092.
40. Pareto made clear, in a very explicit and sometimes pungent manner, the importance of the category of the need for uniformity residues, ibid., §§1115–1132, and of the category of sentiments of hierarchy, ibid., §§1153–1162.

41. Pareto, ibid., §1207, noted that this class of residues "is of the same nature" as interests, but the latter, owing to their great "intrinsic importance for social equilibrium", are better considered separately.
42. Ibid., §888.
43. See (Pareto 1909, chapter II, §19).
44. Ibid., chapter II, §84.
45. Ibid., chapter II, §88.
46. Ibid.
47. See (Pareto 1900, reprinted in Pareto 1980, p. 198).
48. Pareto to Pantaleoni, 17th May 1908, see (Pareto 1984a, p. 103).
49. See Pareto (1917–1919, §1691).
50. Ibid., §1693.
51. Ibid., §1699.
52. Ibid., §1700.
53. Ibid., §1705.
54. Ibid., §1712.
55. Ibid., §§1713–1716.
56. Ibid., §1717.
57. Ibid., §1733.
58. Ibid., §1735.
59. Ibid., §1737.
60. Ibid., §1745.
61. See (Pareto 1901, p. 7).
62. Ibid., p. 8.
63. See (Pareto 1909, chapter II, §102).
64. Ibid., chapter II, §103.
65. See (Pareto 1906, chapter VII, §2).
66. Specifically, due to a surplus of deaths over births and to degenerative phenomena, see (Pareto 1901, pp. 9–10).
67. See (Pareto 1909, chapter II, §103). Pareto repeatedly (Pareto to Giuseppe Prezzolini, 17th December 1903, see (Pareto 1975c, p. 506); to Carlo Placci, 4th January 1904, ibid., p. 513; to Alceste Antonucci, 16th March 1908, ibid., p. 627) rejected outright the accusation of plagiarism made against him by Gaetano Mosca, see (Mosca 1903, p. 200) for having mentioned the theory of élites in his *Socialist Systems* without acknowledging Mosca's recent studies on the topic. Thus, Pareto underlined with transparent sarcasm, Pareto to Alceste Antonucci, 16th March 1908, see (Pareto 1975c, p. 627), that it had long been generally recognised that "it is always a minority that governs, human society is not homogeneous and elites degenerate". He, unlike Mosca, "had conjoined [these facts and] had looked for the interconnections, so that a theory took shape which [he] had checked against the historical data".

68. Pareto underlined that "the political elite is far from being a moral or even an intellectual elite", Pareto to Maurice Millioud, 5th November 1902, Banca Popolare di Sondrio (BPS-la).
69. See (Pareto 1909, chapter II, §104).
70. In Pareto's opinion, in his own period the Trade Unionists had the capacity to overthrow the bourgeoisie, ibid., chapter II, §104b.
71. Ibid., chapter II, §104.
72. Ibid., chapter II, §105.
73. Ibid., chapter II, §106.
74. Ibid.
75. Ibid., chapter II, §105.
76. Ibid., chapter II, §106.
77. Ibid.
78. Ibid.
79. Ibid., chapter II, §107.
80. See (Pareto 1917–1919, §2027).
81. Ibid., §2031.
82. Ibid., §2032.
83. Ibid., §2036.
84. Ibid., §2048.
85. Ibid., §2054.
86. Ibid., §2055.
87. See (Pareto 1909, chapter II, §§110–111).
88. See (Pareto 1899, reprinted in Pareto 1980, p. 176).
89. See (Pareto 1909, chapter II, §85).
90. Ibid., chapter VII, §101.
91. By interests Pareto meant "the ensemble of tendencies on the part of individuals and groups, prompted by instinct and by reason, to appropriate those material goods which are of use in life, or mayhap only of comfort, rather than pursuing esteem and honours", see (Pareto 1917–1919, §2009). He added that "this ensemble plays a major role in the determination of social equilibrium", ibid., and that hence "the findings of pure economics constitute an integral and highly-important part of sociology", ibid. §2013.
92. Ibid., §2205.
93. Ibid., §2207.
94. Ibid., §2091.
95. Ibid., §2207.
96. Ibid., §2206.
97. Ibid., §2207.
98. Ibid., §2208.
99. Ibid., §2209.

100. Ibid., §2210.
101. Ibid., §2211.
102. Ibid., §2212.
103. Ibid., §2215.
104. Ibid., §2210.
105. Ibid., §2217.
106. Ibid., §2217.
107. Pareto to Pio Perrone, 25th January, see (Pareto 1989, pp. 688–689).
108. See (Pareto 1920b, reprinted in Pareto 1987b, p. 584).
109. See (Pareto 1921, reprinted in Pareto 1987b, pp. 657–660).
110. See (Pareto 1917–1919, §2232).
111. Ibid., §2227.
112. Ibid., §§2233–2235.
113. Ibid., §§2234–2235.
114. Ibid., §2235. Pareto, ibid., specified however that the dyad speculators-rentiers should not be confused with the dyad revolutionaries-conservatives because on occasion the rentiers unwittingly contribute to revolutions which are to their own ruin.
115. Ibid., §2416.
116. Ibid., §2553.
117. Ibid. One of the first mentions made by Pareto of the political importance of force was in Pareto to Luigi Bodio, 22nd December 1904, see (Pareto 2001, p. 156), where he declared that "the class where this force is found will come to power". Sometime later, Pareto claimed to believe that "only force, guns and cannons will decide who will be the victors and who the vanquished", Pareto to Pantaleoni, 22nd February 1905, see (Pareto 1984b, p. 437). Pareto touched, too, on the idea that "the big mistake of modern times" is to believe in the possibility of governing without the use of force, which is "actually the foundation of any social organisation", see (Pareto 1909, chapter II, §107b),
118. See (Pareto 1917–1919, §2310).
119. Ibid., §2311.
120. See (Pareto 1902, p. 131). Pareto, ibid., p. 132, reiterated that effective defence of society not only required the suppression or the repression or the education of those elements unfit for living in a social setting but also involved preventing them from procreating.
121. Ibid., p. 130.
122. See (Pareto 1906, chapter VII, §106).
123. See (Pareto 1917–1919, §2427).
124. Ibid., §2420.
125. See (Pareto 1905, reprinted in Pareto 1966, p. 260).
126. Ibid., p. 262.

127. Pareto to Giuseppe Jona, 14th October 1903, BPS-la.
128. Pareto to Papafava, 12th June 1905, see (Pareto 1989, pp. 445–446).
129. See (Pareto 1901, p. 86).
130. Ibid., p. 93.
131. Ibid., p. 102. In Pareto's opinion, see Pareto (1917–1919, §2306), governing is easier in periods of prosperity because this allows governments to finance budget deficits through public debt which can then be paid off by making use of future surpluses.
132. Ibid., §2244.
133. In Pareto's view, ibid., §2252, the differing proportions of force and consent "largely originate from differing proportions of feelings and interests".
134. Ibid., §2244.
135. Ibid., §2245.
136. Ibid., §2255.
137. Ibid.
138. Ibid., §2247.
139. Ibid., §2250.
140. Ibid., §2257. According to Pareto, democratic regimes rely more on clienteles than on the armed forces, compared to non-democratic regimes.
141. Ibid., §2258.
142. Ibid., §2274.
143. Ibid., §2277.
144. Ibid., §§2275–2276.
145. Ibid., §2268.
146. Ibid., §2267.
147. Ibid.
148. See (Pareto 1911, reprinted in Pareto 1966, p. 273).
149. See (Pareto 1917–1919, §2277).
150. See (Pareto 1911, reprinted in Pareto 1966, p. 273).
151. Ibid., p. 276.
152. As regard the costs of going to war, Pareto thought these were more acceptable for a people where the residues of the persistence of aggregates predominate and are commended by the government, see (Pareto 1917–1919, §2454).
153. See (Pareto 1911, reprinted in Pareto 1966, pp. 278–279).
154. Ibid., p. 279.
155. See (Pareto 1917–1919, §2312).
156. Ibid., §2313.
157. Ibid.
158. See (Pareto 1920c, reprinted in Pareto 1980, p. 957).
159. Ibid.

160. Ibid.
161. Ibid.
162. Ibid., p. 958.
163. Ibid.
164. See the pioneering and always of interest (Bruni 1996).
165. Pareto to Pasquale Jannaccone, 20th July 1904, BPS-la.
166. Pareto to Jannaccone, 19th September July 1904, BPS-la.
167. Pareto to François Rouge, 25th November 1904, BPS-la; Pareto to Linaker, 26th May 1905, see (Pareto 1975c, p. 547). With Rouge, Pareto specified once again that he could not "take any firm engagement".
168. Which Pareto, Pareto to Rouge, 25th November 1904, BPS-la, conceived as being made up of the following parts of the *Cours*, "with numerous additions": the first chapter of the second book *General principles of social evolution (Principes généraux de l'évolution sociale)*, the two chapters *The income curve (La courbe des revenus)* and *Social physiology (La physiologie sociale)* from the third book "and, possibly", the first chapter of the first book *(Personal capital) (Capitaux personnels)*.
169. Ibid.
170. Ibid. Since Pareto was aware that this small volume "cannot be a very attractive deal for a publisher", he demanded to receive only 20 complimentary copies in return for its publication.
171. Ibid. Pareto asked Rouge for a remuneration of 500 francs per volume (with the exception of the volume on mathematical economics which Pareto would give free of charge), considering this to be "equitable" given that he intended to provide the publisher with a "substantially new work". He was also willing to prepare a "trial" draft of one volume, with a decision on possible further volumes to be taken in the light of its commercial success, Pareto to Rouge, 6th January 1905, BPS-la.
172. Pareto to Rouge, 15th February 1905, BPS-la.
173. Pareto to Guido Sensini, 9th April 1905, see (Pareto 1975c, p. 543).
174. Pareto to Rouge, 30th November 1906, BPS-la.
175. Pareto to Rouge, 6th December 1906, BPS-la.
176. Pareto to Rouge, 12th December 1906, BPS-la.
177. Pareto to Sensini, 8th December 1906, see (Pareto 1975c, pp. 578–579); Pareto to Sensini, 18th February 1907, ibid., p. 590.
178. Pareto to Sensini, 15th December 1906, ibid., pp. 580–581.
179. Pareto to Rouge, 7th June 1907, BPS-la.
180. Which would become the *Treatise on Sociology*.
181. Pareto to Pantaleoni, October 1907, see (Pareto 1984a, p. 67).
182. Ibid., p. 89.
183. Pareto to Rouge, 11th February 1912, BPS-la. On the failed republication of the Cours, see (Bruni 1997).

184. Pareto to Piero Barbera, 1st April 1906, see (Pareto 2001, p. 162).
185. Some months earlier, Pareto had confessed to not having "yet begun writing the volume on sociology" and to fearing that it would take "some time before it [is] ready", Pareto to Sensini, 20th May 1906, see (Pareto 1975c, p. 565).
186. Pareto to Barbera, 18th September 1906, see (Pareto 2001, p. 163). Incidentally, a few days later, Pareto announced to Pantaleoni that he did not think he would accept Barbera's proposal for the publication of a manual of sociology because, while he had derived "great satisfaction from the publication of the two volumes in French", this had not been the case for the recent publication of the *Manual of Political Economy* in Italian, Pareto to Pantaleoni, 20th December 1906, see (Pareto 1984b, p. 467).
187. Pareto to Barbera, 9th January 1907, see (Pareto 2001, pp. 166–167).
188. Pareto to Linaker, 9th January 1907, see (Pareto 1975c, p. 601).
189. Pareto to Pantaleoni, 30th June 1907, see (Pareto 1984a, p. 40).
190. Pareto to Barbera, 28th December 1910, see Pareto 2001, p. 187).
191. Pareto to Barbera, 18th July 1912, ibid., p. 191.
192. Pareto to Barbera, 22nd August 1912, ibid., p. 196.
193. Pareto later expressed his disappointment at the non-publication of the *Treatise* before the war, as "what was in fact prediction, when it was written, now appears as deduction", Pareto to Barbera, 5th August 1914, ibid., p. 247; Pareto to Bodio, 2nd September 1914, ibid., p. 249; Pareto to de Pietri Tonelli, 25th October 1914, see (Pareto 1975a, p. 881); Pareto to Placci, 26th October 1914, ibid., p. 883.
194. Pareto to Pierre Boven, 1st August 1912, ibid., p. 779.
195. Pareto to Boven, 22nd February 1913, in Pareto, ibid., p. 817.
196. Pareto to Gustave Payot, 28th February 1913, ibid., p. 818.
197. Pareto to Boven, 1st February 1913, BPS-la.
198. See (Pareto 1917–1919, p. 1763, the last page printed of the book).

Bibliography

Bernard, Claude. 1878. *La science expérimentale [Empirical science]*. Paris: Baillère.

Bruni, Luigino. 1996. Gli anelli mancanti. La genesi del Trattato di sociologia generale di Pareto alla luce di lettere e manoscritti inediti [The missing links. The genesis of Pareto's Treatise on general sociology in the light of letters and unpublished manuscripts]. *Il pensiero economico italiano*, pp. 93–135.

———. 1997. The unwritten "Second edition" of Pareto's Cours d'economie politique and its italian translation. *History of Economic Ideas*, pp. 103–126.

Busino, Giovanni. 2008. La science sociale de Vilfredo Pareto [The social science of Vilfredo Pareto]. *Revue européenne des sciences sociales* 140: 107–132.

Mornati, Fiorenzo. 2018a. *An intellectual biography of Vilfredo Pareto, I, from science to liberty (1848–1891)*. London: Palgrave Macmillan.

———. 2018b. *An intellectual biography of Vilfredo Pareto, II, illusions and delusions of liberty (1891–1898)*. London: Palgrave Macmillan.

Mosca, Gaetano. 1903. Il principio aristocratico e il democratico nel passato e nell'avvenire [The aristocratic and democratic principles in the past and in the future]. *Riforma sociale*, anno X, vol. XIII, 15th March.

Pareto, Vilfredo. 1898. *Comment se pose le problème de l'économie pure [Expounding the pure economics]*. Lausanne: Self-Publication.

———. 1899. I problemi della sociologia [The problems of sociology]. *Rivista italiana di sociologia*, March, pp. 145–157,

———. 1900. Un'applicazione di teorie sociologiche [An application of sociological theories]. *Rivista italiana di sociologia*, July, pp. 401–456.

———. 1901. *Les systèmes socialistes [Socialist systems]*, tome I. Paris: Giard et Brière.

———. 1902. *Les systèmes socialistes [Socialist systems]*, tome II. Paris: Giard et Brière.

———. 1903. Proemio. In *Biblioteca di storia economica*, VI–XIV. Milan: Società editrice libraria.

———. 1905. L'individuel et le social [The individual and society]. In *International congress of philosophy. 2nd session held in Geneva from 4th to 8th September 1904. Reports and minutes published under the direction of Dr. Edouard Claparède, secretary-general of the Congress.* Geneva: Kundig, pp. 125–131 and 137–139.

———. 1906. *Manuale d'economia politica con una introduzione alla scienza sociale [Manual of political economy with an introduction to social science]*. Milan: Società Editrice Libraria.

———. 1909. *Manuel d'économie politique [Manual of political economy]*. Paris: Giard et Brière.

———. 1910. Le azioni non logiche [Non-logical actions]. *Rivista italiana di sociologia*, May–August, pp. 305–354.

———. 1911. Rentiers et speculateurs [Rentiers and speculators]. *L'Indépendance*, 1st May, pp. 157–166.

———. 1917–1919. *Traité de sociologie générale [Treatise on general sociology]*. Lausanne-Paris: Payot.

———. 1920a. Discours de M. Vilfredo Pareto [Address by Mr. Vilfredo Pareto]. In University of Lausanne, ed. *Jubilé du professeur Vilfredo Pareto, 1917 [Jubilee of Professor Vilfredo Pareto, 1917]*, 51–57. Lausanne: Imprimerie Vaudoise.

———. 1920b. Un grande dibattito sulle condizioni ed i bisogni dell'industria italiana. Vilfredo Pareto risponde a Pio Perrone [A great debate on the condition and the needs of Italian industry. Vilfredo Pareto replies to Pio Perrone]. *Rassegna italiana*, 29th February, pp. 317–323.

———. 1920c. Trasformazioni della democrazia. III. Il ciclo plutocratico [Transformations of democracy. III. The plutocratic cycle]. *La Rivista di Milano*, 5 luglio, pp. 164–170.
———. 1921. Il protezionismo e i prezzi [Protectionism and prices]. *Il Resto del Carlino*, 7th January.
———. 1966. *Mythes et Idéologies [Myths and ideologies]*. Complete Works, tome VI, ed. Giovanni Busino. Geneva: Droz.
———. 1975a. *Epistolario 1890–1923 [Correspondence, 1890–1923]*. Complete works, tome XIX-2, ed. Giovanni Busino. Geneva: Droz.
———. 1975b. *Jubilé du Professeur V. Pareto [Jubilee of Professor Vilfredo Pareto]*. Complete Works, tome XX. Geneva: Droz.
———. 1975c. *Epistolario 1890–1923 [Correspondence, 1890–1923]*. Complete works, tome XIX-1, ed. Giovanni Busino. Geneva: Droz.
———. 1980. *Écrits sociologiques mineurs [Minor sociological writings]*. Complete Works, tome XXII, ed. Giovanni Busino. Geneva: Droz.
———. 1984a. *Lettere a Maffeo Pantaleoni 1907–1923 [Letters to Maffeo Pantaleoni 1907–1923]*. Complete works, tome XXVIII.III, ed. Gabriele De Rosa. Geneva: Droz.
———. 1984b. *Lettere a Maffeo Pantaleoni 1897–1906 [Letters to Maffeo Pantaleoni 1897–1906]*. Complete works, tome XXVIII.II, ed. Gabriele De Rosa. Geneva: Droz.
———. 1987a. *Marxisme et économie pure [Marxisme and pure economics]*. Complete works, tome IX, ed. Giovanni Busino. Geneva: Droz.
———. 1987b. *Écrits politiques. Reazione, Libertà, Fascismo, 1896–1923 [Political writings. Reaction, Liberty, Fascism, 1896–1923]*, Complete works, tome XVIII, ed. Giovanni Busino. Geneva: Droz.
———. 1989. *Lettres et Correspondances [Letters and correspondence]*. Complete works, tome XXX, ed. Giovanni Busino. Geneva: Droz.
———. 2001. *Nouvelles Lettres 1870–1923 [New Letters 1870–1923]*. Complete Works, tome XXXI, ed. Fiorenzo Mornati. Geneva: Droz.

CHAPTER 6

The War Seen from Céligny

Despite the fact that in this period what Pareto wanted to write was not permitted or was not prudent, particularly in regard to the situation in Italy, while what was permitted or prudent he did not choose to write,[1] he paid close attention to the world war as offering the first significant opportunity to put the tools of sociological analysis he had developed in the *Treatise on General Sociology* to the test.[2] Here we will examine his perspective on war in connection with the international (Sect. 6.1) and with the Italian contexts (Sect. 6.2).

6.1 International and General Aspects of the Conflict

A few weeks after the outbreak of the war,[3] Pareto declared[4] that its principal causes were: "the conflict between German and Slavic nationalism; the conflict between aristocratic militarism and social democracy;[5] the narrow interests of the various nations".[6]

In the first place, the Germanic and the Slavic peoples possessed "a great expansionist drive", which, instead, was diminished in the Anglo-Saxons and had long disappeared in the Latins.[7] In particular, the Germans "are aiming for hegemony in Europe", the Slavs "want to unite into a single entity" and the Anglo-Saxons "want to keep the British Empire intact". However, Pareto considered that while "the clash between the

F. Mornati, *Vilfredo Pareto: An Intellectual Biography Volume III*,
Palgrave Studies in the History of Economic Thought,
https://doi.org/10.1007/978-3-030-57757-5_6

Germans and the Slavs was predestined, inevitable", the one between the Germans and the English could have been avoided, had the Germans not decided to attack "all their enemies simultaneously".[8]

Secondly, conflict between Western Europe and the Central Powers was also "inevitable"[9] because, for the Western European nations, the high levels of military expenditure required to counter German militarism constituted an insurmountable obstacle to the continued increase in social spending which these nations were pursuing in the wake of their conversion to social democracy.[10]

Pareto thus maintained, on the basis of the two factors identified above, that "lasting peace" could be achieved only with the total defeat of one side, making it "likely that this war will be long-lasting", especially as the parties involved "command immense economic resources".

Instead, were only the national interests of the parties at stake, "lasting peace would soon be possible, because these interests are not irreconcilable".[11]

Having observed that the peace bringing an end to the conflict "will [in any case] be precarious, like that of 1870",[12] Pareto added that while "a German victory [would] temporarily halt the plutocratic and demagogic decay of the Latins and now also of the English",[13] "a victory of the Anglo-French alliance would represent, at least in part, a victory for the demagogic plutocracy".[14]

In the Spring of 1915, Pareto extended this analysis, declaring that "when peoples anxious to expand their dominion meet, then conflict is, if not inevitable then at least highly probable".[15] For this reason, the ongoing war appeared to him not to spring directly from economic disputes but from the repercussions of these on the nationalist pride of the peoples involved.[16] In the current conflict, "possibly the most intense" imperialistic impulse was associated with the Germans, taking the form in the English mainly of opposition to the other parties' ventures,[17] while in the Latin peoples it was wholly absent.[18]

On this basis, since the factors which had led to the war were "deep-seated", it would be illusory to imagine that it might forestall other conflicts.[19] Consequently, it could be anticipated that the war would finish either with a peace in which the adversaries "are even", or "with complete, utter and absolute victory for one of the sides". It was clear, however, that in the first case "the peace will only amount to a truce", while in the second case it was difficult to imagine either how "the Triple *Entente* can render Germany impotent for the foreseeable future", thus preventing any attempted reaction, or how "the Central Powers could destroy the immense British or the far-flung Russian empires and stop them from preparing a retaliation".[20]

In any case Pareto considered that victory would go to the not-yet-identifiable[21] alliance in which "strength, discipline and the will to sacrifice the present to the future" would prevail.[22]

In the summer and autumn of 1916, however, Pareto began to sense that "the demagogic plutocracy will probably emerge victorious from the war",[23] especially as "in the march to victory, money and industrial might become as, or even more, important than the efforts of the people or the talents of the commanders": so, the demagogic plutocracy needed no longer fear being stripped of power by "some victorious general".[24] Nevertheless, at the time of the Bolshevik revolution in October 1917, Pareto revised his expectations with the new formulation that "two contrasting forces are now principally at play, nationalist (imperialist) and socialist (subversive)", adding his opinion that, were the war to culminate "with no crushing victory", then the socialist forces would prevail.[25]

In April 1918, Pareto stated his view that the war could still end not only with the victory of one of the two warring parties but also, alternatively, through a "deal between the two opposing sides".[26] In any case, after pointing out, with some sarcasm, that President Wilson, after assigning to his country the mission of "defending America as a whole against European intervention", adding the phrase "of directing Europe, imposing order, justice, democracy and the rule of law",[27] Pareto claimed now to see the war in terms of "a struggle for hegemony between the Germans and the Anglo-Americans"[28] and to be uncertain "which of the two is preferable".[29] Soon afterwards, he added that "political liberty in Europe and the independence of the smaller nations" required a failure on the part of either side to claim a clear victory,[30] expressing his hope that, likewise, "peace should not be imposed in such a way as to allow no reconciliation between the victors and the vanquished".[31] In the early autumn of 1918, Pareto, while considering that "in all likelihood, the military plutocracy is beaten", immediately asked himself whether "the pax Americana will bring the same benefits to the world as the pax romana".[32]

With the ending of the war, Pareto embarked on new deliberations regarding the reasons for its outbreak and for its outcome.

Further scrutinising and partly modifying the original conviction described above, Pareto determined that the conflict "sprang largely out of rivalry between plutocracies" which, exploiting militaristic and patriotic religious sentiment, had for years been cashing in on "the ample profits deriving from preparations for war".[33] In the summer of 1914, signally "by virtue, inter alia, of a period of economic depression, nationalism, imperialism and militarism joined hands with the demagogic plutocracy",[34] with the consequence that "a number of the speculators wanted war",[35] seeing it as "an opportunity for fabulous profits and rewards", while underestimating the risk of defeat and overestimating their own ability to handle victory, hiding "their rapacity beneath the cloak of the democratic and humanitarian mantras, while at the same time attempting to add legal ones".[36]

The socialists, meanwhile, had been against the war,[37] "[both] on the basis of their pacifist ideology and, more particularly, out of fear of militarism". Nevertheless, "they likewise shared in the creed of patriotism and thus never made serious efforts to counteract it".[38]

Pareto then commented that the plutocracies in France and Italy had arrived unprepared for the conflict and in Germany had underestimated the need for adequate diplomatic groundwork, while failing to realise in Russia that war might bring revolution. The only plutocracies which had avoided "miscalculations" had been in America and, "possibly", in England.[39] Then, since in the conflict money (i.e. armaments) rapidly showed itself to be more important than "militias", the plutocracy (particularly that of the Allies) began to view "the end of the war, the source of such good earnings, as a misfortune", becoming ever "more eager to prolong it".[40] Furthermore, according to Pareto, the war "brought the greatest benefit (contrary to what is claimed, as, in the absence of militarism, "the acquiescence of the mob remains the sole mainstay of governments"[41]) to the agricultural and industrial workforce, who were able to earn vast and unhoped-for salaries and to increase their social and political influence, together with newcomers attracted by quick profits, while those who were not prepared or able to participate in these speculations, such as savers, rentiers and the petit bourgeoisie, were adversely affected".[42]

As regards the outcome of the war, Pareto ascribed it to the prevalence of residues of persistence among the rulers of the Central Powers and also of Russia, and to the prevalence of the instinct for combinations among their counterparts in the Alliance.[43] More specifically, Pareto thought that the defeat of the Central Powers was due to "the enormous disparity in human and financial resources, the new characteristics of modern warfare which favoured these aspects and the control of the seas on the part of England and later the USA",[44] with these factors in turn depending on

Germany's appalling diplomatic groundwork in the lead-up to the conflict, omitting to dissuade at least some of its potential adversaries from joining in[45] out of an excessive confidence in its own "destiny, military power, organisation and vital interests".[46]

6.2 The Case of Italy

Regarding Italy's failure to enter the war immediately,[47] Pareto pointed out that Italy, a Latin country led by a radical government, could not support Austria, with its conservative government. However, he also expressed the view that "over and above these considerations, there are also immediate interests with regard to the Adriatic and other matters", which in his view would come to preponderate in Italy's final decisions.[48] Thus, an unspecified "demagogic" party appears to him to favour war against Austria,[49] while the socialist party, although also favouring war with Austria, was "inclined towards neutrality" because it feared that any type of war "would channel strength to the military movement and away from subversive activities".[50]

In the same period Pareto "fear[ed] that Italy possessed too little artillery and was not minimally prepared for war",[51] due to "sympathies and interests" disguised behind the various humanitarian and pacifist ideological positions.[52] More specifically, the Italian government, during the recent war in Libya, had made the mistake of attending "only to [its] interests", thus missing the opportunity to prepare the country for the inevitable world war through an appropriate stirring of patriotic fervour.[53]

Thus, in April 1915, Pareto claimed that "Italy is now at a strange crossroads; either it will come to great fortune or to ruin", adding that this vacillation would endear it neither to the Central Powers nor to the Allies.[54] Even after it had entered the war against Austria-Hungary, on 24 May 1915, Pareto observed that the Italian conduct was still not free of ambiguity in the far-from-secondary question of the decision not to immediately declare war likewise on Germany.[55]

In September 1919, Pareto remarked that Italy, "following its manly and successful conduct of the war",[56] had been met with international hostility, fomented, in his view, by the powers of the United States and of England, who were fearful "of Italy's no longer submitting to the role of subservient handmaiden".[57] However, in April 1920 Pareto observed that in post-war Italy, feelings were gravely agitated by the fact that the war had ended "with political, social and economic disappointments" such as the country's diminished independence, the English command of the Mediterranean and the danger of running out of food and raw materials, as well as of becoming embroiled in further conflicts due to the fragility "of the League of Nations agreements".[58]

Notes

1. Pareto to Napoleone Colajanni, 7th January 1916, see (Pareto 1989, p. 559).
2. In May 1915 Pareto had planned, once the war was finished, to publish an appendix to his *Sociology* to examine "which of the arguments in the book are confirmed [by the hostilities] and which are contradicted, and in this case, wherein lies the error", Pareto to Alfonso de Pietri Tonelli, 21st May 1915, see (Pareto 1975, p. 895). However, this project was abandoned in March 1918 "for innumerable reasons", Pareto to Vittore Pansini, 15th March 1918, see (Pareto 1989, p. 536).
3. An interesting feature of international studies occasioned by the centenary of the First World War is the website www.1914-1918-online.net of the *International Encyclopaedia of the First World War.*
4. Confirming and elaborating on the early impressions reported to Pierre Boven, 27th August 1914, see (Pareto 1975, p. 880). In the same period, Pareto claimed that he had "always known for sure" about this war, even if he had not been in a position to foresee "when it would occur", Pareto to Luigi Bodio, 2nd September 1914, see (Pareto 2001, p. 249).
5. Pareto also emphasised that "this war destroys the illusion … [of the] pacifists who were trying to lead us to believe that the general laws of human evolution had changed", Pareto to Bodio, 28th December 1915, in Pareto, ibid., p. 269.
6. See (Pareto 1914a, reprinted in Pareto 1987, p. 523).
7. Ibid., p. 524.
8. Ibid., p. 525.
9. Ibid., On the other hand, this view was later tempered. Initially, Pareto stated as his opinion that the conflict could have been "supplanted by civil wars" in the countries concerned, Pareto to Bodio, 2nd September 1914, see (Pareto 2001, p. 250) and 6th November 1915, ibid., p. 267. Then, a few months later, he maintained that had "democrats and plutocrats been less greedy for public funds for their own use and for electoral purposes", thereby allowing France and England to make better preparations for war, Germany might not have attacked them, see (Pareto 1915a, reprinted in Pareto 1920a, p. 47).

10. See (Pareto 1914a, reprinted in Pareto 1987, pp. 525–526). On the other hand, in keeping with his most recent views on the subject prior to the war, see above §4.8, Pareto repeated that "the plutocracy did not want war but unwittingly facilitated it by promoting, through its newspapers, the hardening of sentiments of hatred between the nations" which then joined in war, see (Pareto 1914a, reprinted in Pareto 1987, p. 527). He later scaled down this assertion, speaking simply of a school of thought, always a minority within the plutocracies of the nations of the Entente, which had been against the war, considering it "better to be content with what is little but certain than to run, overcome by blind cupidity, after what is greater but more uncertain", see (Pareto 1920a, pp. 366–368).
11. See (Pareto 1914a, reprinted in Pareto 1987, pp. 525–526).
12. See Pareto to Carlo Placci, 26th October 1914, see (Pareto 1975, p. 882).
13. Pareto to Bodio, 2nd September 1914, see (Pareto 2001, p. 250).
14. Pareto to Bodio, 6th September 1914, ibid., p. 252. In October 1914 and in October 1915 Pareto identified a corroboration for his contention that it was sentiment and interests which determined people's thought processes, and not the other way around, in the fact that the accusations against the Germans for their violation of Belgian neutrality had been accepted *a priori* by Germany's enemies and rejected, again *a priori*, by their supporters, see (Pareto 1914b, reprinted in Pareto 1987, p. 532); Pareto to James Harvey Rogers 22nd October 1915, see (Pareto 1975, p. 904).
15. See (Pareto 1915a, reprinted in Pareto 1920a, p. 37).
16. Ibid.
17. Ibid.
18. Ibid., p. 38.
19. Ibid., p. 53.
20. Ibid., p. 55.
21. Pareto to Guido Sensini, 6th November 1915, see (Pareto 1975, p. 906). In general terms Pareto thought that the French and English governments of the time were pursuing the dual aims of, firstly, "saving their demagogic plutocratic party from the consequences of its negligence in spending the money which should have served to prepare for war on elections and, secondly, saving the nation from being exploited by this party". Since these aims do not coincide perfectly, the war efforts of these governments are less effective than those of the German government which was pursuing the "almost interchangeable" aims of "preserving militarism and bringing victory to the nation", Pareto to Bodio, 6th November 1915, see (Pareto 2001, p. 268). In October 1915, Pareto claimed that "it is not in terms of organisation that Germany has placed itself above the allies, it is in the sentiment whereby people are induced to accept individual sacrifices for

the good of the country and whereby they are prevented from disregarding the future in order to enjoy the present", see (Pareto 1915b, reprinted in Pareto 1966, p. 287). In any case, in Pareto's view, in the absence of the Russian threat, if Germany had waited another 20 years, "with little effort it would have overcome the plutocratic-demagogic peoples, who would have fallen into its lap like a ripe pear from the tree", Pareto to Bodio, 2nd September 1914, see (Pareto 2001, p. 250) and 5th April 1915, p. 258, ibid.

22. Ibid. On the other hand, in this period Pareto doubted whether it was yet possible to "ascertain" whether the war was harmful or otherwise, bearing in mind that, in general, terrible wars had been followed by advances in civilisation and that in the conflict which had just begun, material and human losses were indeed perhaps greater than in earlier conflicts in absolute terms but possibly were lower in relative terms, that is in relation to the population and to material resources, Pareto to Bodio, 31st March 1915, see (Pareto 2001, p. 257), and 6th November 1915, ibid., p. 267.
23. Pareto to René Johannet, 19th July 1916 (Banca Popolare di Sondrio: Vilfredo Pareto's letters archive) BPS-la; Pareto to Arcangelo Ghisleri, 4th August 1916, see (Pareto 1975, p. 931).
24. See (Pareto 1918a, reprinted in Pareto 2005, p. 235). On this fear as the reason for the speculators' vacillation in regard to the war on the eve of the conflict, see above, §5.4.
25. Pareto to Pansini, 20th October 1917, see (Pareto 1975, p. 620). In the spring of 1918, Pareto wondered whether the fact that "as things stand, the working class has benefitted most from the war, other things being equal" had anything to do with its "abandoning of the principles of Marxist internationalism" and its conversion "to patriotic imperialism", see (Pareto 1918b, reprinted in Pareto 1967, p. 41).
26. Ibid., p. 25.
27. Ibid., p. 33. The United States had joined the war on 2nd April 1917. In May 1919, Pareto, remembering that in July 1918 "[I was] almost the only one not to prostrate [my]self before Wilson [and] was one of the very few in Italy to perceive the threat of Anglo-American hegemony", noted that "now" most Italians "suffer from the American domination", Pareto to Bodio, 17th May 1919, see (Pareto 2001, p. 327).
28. See (Pareto 1918c, reprinted in Pareto 1920a, p. 154).
29. See (Pareto 1918b, reprinted in Pareto 1967, p. 28). Shortly after the beginning of the conflict, Pareto had claimed that "[his] sentiments" led him to "hope that the Germans (who are truly odious)" would not win, Pareto to Bodio, 5th April 1915, see (Pareto 2001, p. 258) ibid., and to Moneta, 28th December 1915, ibid., p. 271.
30. See (Pareto 1918b, reprinted in Pareto 1967, p. 79).
31. See (Pareto 1918c, reprinted in Pareto 1920a, p. 168).

32. Pareto to Pantaleoni, 11th October 1918, see (Pareto 1984, p. 239).
33. See (Pareto 1920a, p. 358).
34. Ibid., p. 359.
35. Ibid., p. 358, note 2.
36. Ibid., pp. 366–368.
37. Ibid., p. 360, note 1.
38. Ibid. In the spring of 1919, Pareto claimed that the socialists within the Central Powers supported the war "overcome by greed for the spoils they hoped to win through an alliance with the bourgeois parties", see (Pareto 1919a, reprinted in Pareto 1920a, p. 230).
39. See (Pareto 1920b, reprinted in Pareto 1980, p. 960).
40. See (Pareto 1920a, p. 362).
41. Ibid., p. 366.
42. Ibid., p. 365.
43. Ibid., p. 346; Pareto to Giuseppe De Flamini, 6th February 1919, BPS-la. In Pareto's view, there was likewise an abundance of persistence of aggregates among the citizens of the warring nations, see (Pareto 1920a, p. 347).
44. Ibid., p. 336.
45. Ibid., p. 337.
46. Ibid., pp. 345–346. On the other hand, in March 1918 Pareto was still able to affirm that "even if Germany is defeated, the fact of having been able to resist for so long against the forces of the whole world demonstrates once again the value of a governing class rich in first-class residues and of a population rich in second-class residues", Pareto to Pansini, 15th March and 20th December 1918, see (Pareto 1989, pp. 636, 650).
47. On 3rd August 1914, Italy, despite having been a member of the Triple Alliance together with the German and Austro-Hungarian Empires since 1882, had declared its neutrality on the basis that Austria in entering the war had not triggered the *casus foederis.*
48. Pareto to Boven, 27th August 1914, see (Pareto 1975, p. 880).
49. Pareto to Bodio, 6th September 1914, see (Pareto 2001, p. 252).
50. Ibid.
51. Pareto to Pantaleoni, 7th October 1914, see (Pareto 1984, p. 175).
52. See (Pareto 1914, reprinted in Pareto 1987, pp. 530–531).
53. See (Pareto 1914c, reprinted in Pareto 1987, p. 538). From the perspective of the Italian economy, Pareto expected that were Italy "to remain neutral and to remove the senseless bans on exports, it will make money, just as happened after 1870". Instead, if it entered the war, it would suffer no harm as long as the conflict was short and victorious, while an extended conflict, even if victorious, "would lead to a serious economic crisis", Pareto to Pantaleoni, 19th August 1914, see (Pareto 1984, p. 173).

54. Pareto to Pantaleoni, 15th April 1915, ibid., p. 176.
55. Pareto to Harvey Rogers, 22nd October 1915, see (Pareto 1975, p. 905). Italy declared war on Germany only on 28th August 2016.
56. See (Pareto 1919b, reprinted in Pareto 1920a, pp. 277–278).
57. Ibid., p. 279.
58. See (Pareto 1920c, reprinted in Pareto 1987, p. 622).

Bibliography

Pareto, Vilfedo. 1914a. Conflitto di razze, di religioni e di nazioni [Racial, religious and national conflict]. *Il Giornale d'Italia*, September 25.

———. 1914b. «Invece di provvedere artiglierie ed armi si spendevano i quattrini per fini elettorali». Ma gli Stati si difendono con le armi e non con le chiacchere [«Instead of replenishing artillery and armaments money was spent for electoral ends», while national defence requires arms, not prattle]. *Il Giornale d'Italia*, October 15.

———. 1914c. Non si ottengono vantaggi senza sacrifice [Advantages cannot be gained without sacrifice]. *Il Giornale d'Italia*, November 16.

———. 1915a. La Guerra ed i suoi principali fattori sociologici [The war and its principal sociological factors]. *Scientia*, March, pp. 257–275.

———. 1915b. L'Allemagne a-t-elle le secret de l'organisation ? Réponse de M. Vilfredo Pareto [Does Germany possess the secret of organisation ? Vilfredo Pareto's answer]. *L'opinion*, October 16, pp. 295–296.

———. 1918a. Les conséquences probables de la guerre [The probable consequences of the war]. Céligny: unpublished manuscript.

———. 1918b. *Mon journal [My journal]*. Céligny: unpublished manuscript.

———. 1918c. Après quatre ans de guerre [After four years of war]. *Coenobium*, July–August, pp. 1–24.

———. 1919a. Speranze e disinganni [Hopes and disillusionment]. *Rivista d'Italia*, marzo 31.

———. 1919b. *Realtà [Reality]*. *L'Idea Nazionale*, September 7.

———. 1920a. *Fatti e Teorie [Facts and theories]*. Florence: Vallecchi.

———. 1920b. Trasformazioni della democrazia III. Il ciclo plutocratico [Transformations of democracy III. The plutocratic cycle]. *Rivista di Milano*, July 5, pp. 164–170.

———. 1920c. I provvedimenti finanziari. Un problema insolubile [The financial provisions. An insoluble problem]. *Il Resto del Carlino*, April 28.

———. 1966. *Mythes et Idéologies [Myths and ideologies]*. Complete Works, tome VI, ed. Giovanni Busino. Geneva: Droz.

———. 1967. *Sommaire du cours de sociologie suivi de mon journal [Summary of the course in sociology followed by my journal]*. Complete Works, tome XI, ed. Giovanni Busino. Geneva: Droz.

———. 1975. *Correspondence, 1890–1923 [Epistolario 1890–1923]*. Complete Works, tome XIX-2, ed. Giovanni Busino. Geneva: Droz.

———. 1980. *Écrits sociologiques mineurs [Minor sociological writings]*. Complete Works, tome XXII, ed. Giovanni Busino. Geneva: Droz.

———. 1984. *Lettere a Maffeo Pantaleoni 1907–1923 [Letters to Maffeo Pantaleoni 1907–1923]*. Complete Works, tome XXVIII.III, ed. Gabriele De Rosa. Geneva: Droz.

———. 1987. *Écrits politiques. Reazione, Libertà, Fascismo, 1896–1923 [Political writings. Reaction, Liberty, Fascism, 1896–1923]*. Complete Works, tome XVIII, ed. Giovanni Busino. Geneva: Droz.

———. 1989. *Lettres et Correspondances [Letters and correspondence]*. Complete Works, tome XXX, ed. Giovanni Busino. Geneva: Droz.

———. 2001. *Nouvelles Lettres 1870–1923 [New Letters 1870–1923]*. Complete Works, tome XXXI, ed. Fiorenzo Mornati. Geneva: Droz.

———. 2005. *Inédits et Addenda [Unpublished writings and Addenda]*. Complete Works, tome XXXII, ed. Fiorenzo Mornati. Geneva: Droz.

CHAPTER 7

The Post-war Period

Pareto's wide-ranging and thought-provoking analysis of the post-war period[1] started out from the idea that from the moment "the demagogic plutocracy had achieved a complete victory" in the recent war, "the battle between this force and out-and-out demagogy"[2] had begun, with no means of foreseeing the outcome. This new confrontation was closely monitored by Pareto, with continuing reference to the conceptual tools developed in the *Treatise on General Sociology*, with the result, in his view, to be decided once again by "force".[3] In the following sections we will reconstruct the phases of his investigations, divided into the following categories: a non-economic interpretation of social conflict (Sect. 7.1); reflections on the problems besetting the League of Nations (Sect. 7.2); his views on the early phases of the Bolshevik revolution in Russia and the attempted power grab in Italy in the summer of 1920 (Sect. 7.3); his views on the early rise of Fascism (Sect. 7.4) and his parallel analysis of the principal economic issues of the post-war period (Sect. 7.5).

7.1 A Non-economic Interpretation of Social Conflict

In May 1920 Pareto observed that "in all human communities there are two contrasting forces at work, one attempting to concentrate power centrally and the other tending to disperse it". These forces in turn were fed

F. Mornati, *Vilfredo Pareto: An Intellectual Biography Volume III*,
Palgrave Studies in the History of Economic Thought,
https://doi.org/10.1007/978-3-030-57757-5_7

by residues of the persistence of aggregates and by residues associated with sociality.[4]

When the centrifugal force predominates, the power of disparate individuals and groups is increased, giving them increasing independence and causing those who feel under threat from this to look "elsewhere for protection, no longer shielded by a centralised power, but swearing allegiance to a powerful figure or banding together among themselves".[5] In contrast, the centripetal force is associated mainly with the religious sentiments emerging from the residues of the persistence of aggregates or alternatively with victories in international conflicts[6] (while defeats, or even simply the demands of a very challenging war on the population involved, would favour the centrifugal force).[7] At this time, only "the interests of the plutocrats and the state religion witnessed in the nationalists or in the pseudo-Marxist or anti-Union socialists"[8] bolstered the central power. An interesting example of the oscillations in the equilibrium between the centrifugal and centripetal forces is provided by contemporary English political history where parliament, the dominant force in the early nineteenth century, was obliged after the First World War to deal with the unions on equal terms.[9]

7.2 An Ineffectual New International Organisation and an Unenforceable Peace Treaty

As early as March 1919, Pareto foresaw that the League of Nations would be a further example of "material appetites dressed up in fine idealistic garb",[10] since "the basis of the [new body] is the hegemony of England and of the United States",[11] countries obliged to maintain "powerful armies"[12] in order to match their imperialistic ambitions, which would nevertheless be difficult to achieve because "sooner rather than later the partnership will degenerate into open rivalry".[13] Thus, in February 1920, Pareto noted with evident gratification that the Republican party in the United States Senate had spurned "the League of Nations, not wishing to become entangled in that hotbed of wrangling",[14] which incidentally was fuelled by the "so-called principle of nationality [which], far from smoothing out past conflicts, is generating new ones".[15] The plutocracy, on the other hand, was in favour of the League of Nations, in the hope that, for example, "it might place whole swathes of worldwide economic activity within the grasp of the Trusts".[16] Pareto considered that in general "the plutocracy is instinctively against [separate] states [and] instead works for political centralisation because it knows it can direct this process to its own advantage":[17] hence, although in Germany "separatist feelings are still very strong" the plutocracy numbered among the factors which induced France to avoid insisting on the dismemberment of Germany, despite the fact that this would have ensured "a security not otherwise obtainable".[18]

Pareto's scepticism with regard to these ongoing efforts to create an international political organism[19] was confirmed on the occasion of the international conference held in Cannes from 6 to 13 January 1922 which, as he noted, had attempted to reconcile insoluble political, economic and financial problems,[20] the most important being the avoidance of renewed Franco-German hostilities. Here, in fact, the ideas proposed to induce the Germans to adopt a more conciliatory position (such as the League of Nations or a renewed Anglo-French alliance) were, in Pareto's view, pointless, because "the less a treaty comes over time to agree with the interests of the parties involved, the less it is invoked, whereas if it does agree, it is almost superfluous".[21]

Shortly after this, in view of the new international conference being convened in Genoa,[22] Pareto noted that here, as for the conference in Verona promoted by the Holy Alliance a century earlier, the plutocracy (now "mature and dominant, but at that time just coming into existence") sought to "regulate the affairs of all nations, while maintaining the status quo".[23] At Verona, the nations which had defeated Napoleon had "the sensible idea of including France itself, weakened but not reduced to penury, in their gathering",[24] but deluded themselves that "they could halt the irresistible advance of democratic sentiment", when they could have simply harnessed it "for purposes of government".[25] According to Pareto, where "the Congress of Vienna [thus] marked the beginning of the end for the Holy Alliance", the meeting in Genoa "could foreshadow the end of the current post-war order",[26] in that the "one step which might be of use, the amendment of the treaties which appear increasingly difficult to implement with each passing day",[27] as being "the product of ideology and of blind hatred",[28] was not up for discussion, when, instead, amendments to these might lead to a resolution of the political and social problems and consequently also of the economic problems. Thus, the meeting in Genoa would go ahead in notional terms but would be in vain.[29]

In the following month of May, Pareto reiterated that the treaty of Versailles[30] was misconceived, firstly because it was based on the premise "that it would be possible to keep such a numerous and tenacious people [as the German people] in complete subjugation for thirty years or more" and secondly because it was presumed that it could be easily and cheaply implemented by the victors.[31] He further underlined that reconciliation between Germany and Russia was inevitable,[32] "because history shows that cooperation leads them to victory and disharmony to defeat".[33] Then, in May 1923, Pareto confirmed his view that "for the present, the League of Nations is marking time and allows itself to be led into dealing with trivial matters so as to justify its existence".[34]

In this context, with specific reference to the war reparations which had been formally imposed on the Germans, Pareto underlined the crucial logical distinction between "what Germany should pay, what Germany can pay and what it is in the victors' interest to be paid",[35] stating that in general "knowing the amount of money which is due from a defeated enemy is far less important than knowing the amount he is able to pay".[36] Hence, by March 1922 Pareto had reached the definitive conclusion that "Germany will never pay the amount demanded", the [only] question being whether it doesn't pay "because it does not wish to, or because it can't",[37] adding further that in any case, Germany would not be able to pay the war reparations if it was prevented from returning to economic prosperity.[38]

Hence, France would have to retain its deficit of five billion francs[39] (which it had hoped to finance out of the German war reparations) since it would be difficult to derive any significant economic benefit from its occupation of the Ruhr.[40]

Moreover, in Pareto's view, the non-payment of the German war reparations, taken together with the outstanding English, French and Italian war debts to the United States and with the ongoing "pursuit of capital" would destabilise "a number of countries" economically, paving the way "for the campaigns of violence which have commonly brought an end to problematic situations of this type".[41]

7.3 Bolshevism: The October Revolution, the First Acts of the Bolshevik Regime and the Bolshevik Conspiracy in Italy in the Summer of 1920

Notwithstanding the dearth of information, as early as June 1917 Pareto claimed to be following “the Russian revolution very closely”, in order to “establish whether this sociological experiment will confirm, modify or shatter existing scientific consensus”.[42]

The first and possibly the most important aspect which struck him, prompted initially by the anti-Bolshevik crackdown in the wake of the first Pan-Russian congress held in Petrograd from 16 June to 7 July 1917, was the “confirmation that the stability of a government rests on consent and on force”. Pareto remarked that whereas the provisional government had thought “it could maintain power through consent alone … it has already had to resort to force in jail[ing] the rebels, exactly as under Tsarism”.[43] Having reiterated[44] that the fall of Nicholas II[45] and of Kerensky,[46] as well as the failure of Kornilov’s attempted coup d’état,[47] were due to the lack of army support, Pareto opined that Lenin had “endured” because he was happy and able to use the force.[48] In September 1922 Pareto declared once again that “the Soviet government survives almost entirely thanks to its generalised use of force, which makes up for its critical social failings and for its monumental errors in economics”.[49]

More generally, it was in opposition to the plutocracy that “Bolshevist sentiments first arose”.[50] These sentiments encompassed prioritising “class struggle over political conflicts”,[51] “the promotion of material pleasures above intellectual or emotional ones”,[52] “the abandoning of the goddess democracy whose divinity [the] Bolsheviks deny outright”,[53] showing “disdain for the prevailing religion with its condemnation of the use of force for the retention of power as abhorrent”[54] and “overturning the nature of social relations between classes, [placing] those who were bottom at the top and vice versa”.[55]

On the basis of this, and bearing in mind that “opposing parties may have elements in common which are manifested in different ways”, in Pareto’s view the League of Nations and Bolshevism differ only with regard to those on whom they wish to confer “dominion” over the world, with the former intending to assign it “to the political order of the

Anglo-American alliance and to the social order of the demagogic plutocracy", the latter to "the proletarians of all nations".[56]

Regarding the prospects for Bolshevism, in January 1919, after having over the preceding months characterised "the direct and total repudiation of debts which had just been enacted[57] in Russia[58] as a brutal procedure" and deplored the "benevolence extended by certain humanitarians" to the Bolsheviks, regardless of their "bloodthirsty and thievish deeds",[59] Pareto described his impression that "with multiple transformations, Bolshevism [may] engender a healthy future society".[60] However, as of February 1921 Pareto became convinced that "slowly but surely the Russian system is turning into a red form of Tsarism",[61] that is bellicose, imperialistic[62] and naturally "with the usual opulent, profligate pleasure-seekers at the top, and the plebs living in misery and squalor below".[63]

As to the early economic results of the Bolshevik regime, towards the end of 1921, Pareto, noting that "Lenin is seeking to substitute communism for private property",[64] expressed some doubt whether "the Soviet system [is] compatible with economic activity" since the regime's secret police appeared to impede Lenin's own attempt to seek the "capitalist assistance" which he himself deemed indispensable.[65] In this regard, Pareto was nevertheless of the opinion, subsequently also corroborated by history, that there might exist communists "who, in practice, make use of capitalism". As an example, he stated that there was no difference between "recognising private ownership of a mine or alternatively declaring it communist property and then granting a concession on it for a century, or possibly less".[66] In any case, in January 1923 Pareto gave as his definitive opinion that "the fruits of the equality which a participant in a Russian soviet enjoys are far less than what a skilled worker in the hierarchical society of the United States will be entitled to".[67]

In September 1920, Pareto considered the contemporary occupation of factories in Italy[68] to represent, firstly, a confirmation of his prediction regarding the crumbling of centralised power[69] and secondly, a further example of the confrontation between the *élite* which is in command and another seeking to take its place.[70] This type of confrontation generally follows a pattern whereby at first the group in power will simply seek to cling on to it. Then, in a second phase, it will relinquish some of the trappings of power while keeping control of the substance before, in the final phase, it is overthrown, bringing the beginning of "a new cycle which, [however], to some degree resembles what went before",[71] with the ongoing conflict in Italy corresponding to the second of these phases.[72]

The joint control of factories introduced through government mediation could not last and so either the proprietors would prevail, effectively preserving the former status quo, or the workers would do so, thus "usurping the current owners". Such a development, according to Pareto, "might constitute the lesser evil for society" because, fundamentally, "throughout history, proprietors have often changed while property has remained".[73] Pareto added that such control, if it simply amounted to giving the workers access to the enterprise's true financial situation, would change little with regard to the existing situation, where they were already able to "extract the most they could from the bosses, in the light of market conditions".[74] From the organisational point of view, given that capital would continue to be provided by the private sector,[75] the most interesting prospective aspect was that the socialists conceived of enterprises run by the collective (as distinct from the state bureaucracy) whereas the Trade Unionists imagined them run by the Trade Unions, a form of management which was "untested, whose details would be arrived at through trial and error and experience".[76] In the meantime, a significant fall in production was in any case to be feared,[77] particularly if "management decisions were taken in ignorance of the risks, with a consequent tendency to squander capital".[78]

7.4 Fascism: From Anti-Bolshevik Reaction to Dominant Governing Force

In February 1921, Pareto commented on a significant continuing disparity in Italy, at both the economic level (with symptoms of this being "the high cost of living, unemployment, the oppression of savers, fiscal abuse of the currency, irrecoverable budget deficits") and the social level (with symptoms including "prevarication on the part of private individuals, signs of impending civil war bereft of any kind of idealism, the threat of revolutions ready to sweep away all civic order").[79]

This aside, according to Pareto "the most worrying thing is that there is no perceptible sign of abatement of antisocial feelings".[80] As regards what were the most dynamic political forces, Pareto noted that in fact the ideals of Trade Unionism and of the more extreme brand of socialism were met with "enormous practical difficulties"[81] while the ideals of the Fascists, whose actions appeared "to resemble those of a militant faction rather than of a revolution",[82] remained unclear.

This represented Pareto's first mention of the Fascist phenomenon[83] which, in the absence of a mythology[84] "to make it revolutionary", appeared to Pareto to "be lurching towards a state resembling the white terror which emerged in France at the time of the Restoration". Thus, in his view, invariably "preferable are countries" displaying neither Fascist terror nor "red tyranny", which in any case the events of the summer of 1920 had shown to be no better.[85]

However, on exploring this idea further, Pareto soon came to see Fascism as providing additional and unequivocal confirmation that "when the state does not fulfil its duty of maintaining order and social regulation, the task falls to private citizens".[86] Meanwhile Giolitti, believing that the current social cycle was not yet concluded,[87] sought[88] to restabilise the disturbed social equilibrium by restraining "those who [were] becoming domineering" and bolstering "those who [were] losing strength".[89] Specifically, he had allowed socialism and Trade Unionism to subject the bourgeoisie to such persecution that "Fascism emerged as a response, leaving the government no longer powerless in the face of those wishing to subvert society".[90]

Again in June 1922, however, after having further claimed that the bourgeoisie "accepts the tutelage of the Fascists [and] possibly provides them covertly with financial assistance",[91] Pareto stated that he did not see "in Fascism a profound and durable force".[92] Then, in August 1922, he underlined that, in general, "if those aiming to upset the established order stop halfway after having won a victory, they lose the fruits of their victory and veer towards defeat", just as had happened two years earlier "to the communists in the wake of their occupation of landholdings and factories".[93]

Pareto was in any case of the view that the anarchy currently prevailing in Italy could not last.[94] Then, in the following September, while once again declaring that "either the Fascists will start the revolution now, or never, with the latter possibility now seeming the most likely",[95] he also said that Ettore Ciccotti,[96] "speaking in favour of dictatorship",[97] was expressing "what is contained in many people's hearts, because of the current establishment's inability to resolve the weighty problems which are afflicting it".[98]

However, in October Pareto stated that if Fascism followed Mussolini in his recent antidemocratic stance, "radical changes could ensue"[99] in the Italian political scene, which he thought comparable to that of 1849 in France but without a "Luigi Napoleone Bonaparte; if [such] could be found, a coup d'état would be certain and the outcome beyond dispute".[100] He nevertheless insisted that if the Fascists did not quickly foment the revolution, they would be neutralised by "that old fox Giolitti"[101] and subsequently forsaken by the "multitudes who are even now abandoning the socialists".[102] While even on the eve of the march on Rome he remained unconvinced of "how the Fascists will be able to tackle the formidable financial and economic problems",[103] immediately afterwards he was speculating on Mussolini's ability "to rid himself of the encumbrance of his followers", despite appearing to Pareto "a statesman of exceptional merits".[104] Later, in agreement with the argument advanced by the nationalist journalist Vincenzo Fani whereby the Fascist revolution had come about because, unlike the attempt made by the socialists two years earlier, it was supported by true believers,[105] Pareto reiterated that "greatest effects of social action" are seen when the leaders are plentifully endowed with instincts for combinations and the followers with the persistence of aggregates. Pareto then observed that the socialists had not taken over a decisive role in the state because their militants had preferred to stop with the immediate fulfilment of their specific interests with the occupation of land and factories, salary increases and increases in local taxes to be divided among themselves.[106] On the other hand, the Fascist militants had pursued the ideals of national prestige, centralised authority and abhorrence of democracy and had been guided by a leader who, in order to attain complete control, had not hesitated to refuse the easy option of a leading role in a parliamentary government which had been offered to him.[107]

In any case, after Mussolini's government had taken office,[108] Pareto rapidly concluded that "in Italy things are going well",[109] declaring with unwonted enthusiasm that "the victory of Fascism provides a splendid demonstration of the prognostications of [his] *Sociology* and of many other among [his] writings", an event regarding which he "could be gratified both as a man and as a scientist".[110] However, already by the month of December Pareto reported that "signs are appearing, albeit faintly, of a future less bright than could be hoped". In truth, the use of force, while "normal [when] associated with major and essential issues", became "inappropriate [when] it extends beyond these limits", particularly if performed with a view to "limiting the expression of ideas, even partisan ideas".[111]

In general terms, a dictatorship must necessarily limit freedoms, but then needs to distinguish among and deal appropriately with opponents on the basis of the various levels of threat they pose, if it wishes to establish a long-lasting regime.[112] However, in order to maintain power, Fascism would need, particularly and variously, to: avoid being swallowed up by the other parties;[113] take seriously such constraints as: "the need for the majority to work harder and consume less"; the necessity of doing without "significant state functions in order to make major savings"; the impossibility of "forming a government not controlled by an elite"[114] (as explicitly recognised by Fascism, to its "great merit"); the need to improve the financial position of the population, even slightly, because the masses placed greater importance on their material interests than on their political orientation, fickle as it is.[115]

Pareto, for his part, gave particular prominence to the need for Fascism to move towards a constitutional reform which would bring the institutions, hitherto unchanged, into line with the major innovation ushered in by the Fascist revolution.[116] Since "extended government by force" was not an option, what was required was "at least the tacit consensus of the majority and for this a parliament would be of great service and an extensive freedom of press indispensable".[117]

However, as regards parliament, if "it were retained", it had to be in such a form as to "do some good, with the least damage possible".[118] Thus, the question of the electoral law was much less important than "the powers to be given to it",[119] which should be limited to the voicing of "feelings, interests and even prejudices strictly of a general nature", with technical matters, instead, to be dealt with by "a competent Council of State, together with producers' and consumers' councils".[120] Again, with a view to "depriving the chamber of the power to do damage", further possible steps could include: giving the representatives the sole power of reducing expenditure; drastically reducing their powers of interpellation; giving the government powers "of tax collection and of expenditure on the basis of the previous budget, when a new one has not been approved"; giving "greater powers to the Senate"[121] and making "limited use of referendums".[122] In the last analysis, Pareto hoped that "the appearance of a majority system could be retained, to gratify strong feelings, but the substance [should] go to an *élite*, since [it] is for the best, in objective terms".[123]

Pareto also thought that a government emerging from the Trade Unions, while technically more competent than a political government, would not bring an end either to conflicts between the unions or between these and the rest of the population.[124] Hence, in Pareto's opinion, a political government (whether parliamentary or otherwise), assisted by specialists reporting to it,[125] was always to be preferred, with new laws being prepared by a Council of State with the participation of the Trade Unions.[126]

7.5 The Principal Economic Problems of the Post-war Period

The numerous studies in applied economics published by Pareto in the post-war period appear to gravitate, with multiple interconnections between them and with the matters covered in the preceding sections, around the implications of a generalised excess of consumption over production (Sect. 7.5.1), the problems of public finance (Sect. 7.5.2) and the phenomena of devaluation and of inflation (Sect. 7.5.3).

Macroeconomic Disequilibrium

In January 1922, Pareto repeated once more that "the war was desired by a minority who, in order to motivate the majority to participate, made numbers of promises which could not be maintained, with the dual consequences of provoking the justified anger of those who had been duped as well as distorting the true situation".[127]

On behalf of the Central Powers, these promises had been made by the military plutocracy and, being "principally of a political nature, could easily have been maintained in case of victory". Instead, for the nations of the Alliance the promises were made by the demagogic plutocracy and, being "principally of a social nature", could be maintained only by finding "significant resources",[128] which could be derived neither from an illusory post-war reduction in arms expenditure nor from German war reparations.[129]

Basically, the economic equilibrium prior to the war had been "real rather than notional", based on "a certain number of hours of labour and certain salaries valued in terms of quantities of commodities" while after the war, to favour the workers, "a different system was pursued", with higher real salaries and shorter working hours.[130] It was a question, therefore, of determining whether this new equilibrium "could endure, or not". Pareto inclined to the negative because it would not be possible to "provide" for a marked increase in consumption for the "majority of the population, out of the relatively small quantities of wealth at the disposal of the well-to-do classes and the rich".[131]

Pareto had thought that the difficulties of this transition could nevertheless be attenuated by increases in the production of consumer goods,[132] which instead was hindered, firstly by the unemployment benefits which acted as a disincentive for the conversion to agriculture[133] of those "unable to find employment because of their excessive salary demands",[134] secondly by the licit or illicit occupation of land said to be uncultivated or poorly cultivated, which in turn was a consequence of the fact that "the agricultural workers had been promised the earth and the sky after the war"[135] and lastly "by the haemorrhage of capital intended for production caused by government loans, issues of paper money and taxes" which left those wishing to make productive investments without even "the hope of not being despoiled, given that the domestic situation is even more disturbed than the international situation".[136]

In the absence of unpredicted increases in production and/or reductions in consumption, a return "to the state of equilibrium between labour and capital" existing prior to the war would require a crisis,[137] whose onset Pareto announced in July 1920, claiming to have already anticipated it in 1913, as being the inevitable consequence of the period of prosperity which had begun at the end of the 1800s,[138] a consequence which had simply been delayed by the conflict. The symptom of this crisis, which had begun in Japan, passing then to the United States followed by Germany, England and Italy,[139] was the increase in interest rates, starting from the 7.5% rate on public debt in the United States and then extending to other, perfectly solvent states such as the Swiss Confederation and the Kingdom of Belgium.[140] The factor triggering this increase in interest rates was to be found "in the destruction of capital on the part of governments through additional tax demands and public debt issues, measures which expropriated capital, with only a negligible part being used for production".[141]

The ongoing crisis which was affecting all civilised countries (including those which had not taken part in the war) was a complex phenomenon and so it was absurd to attribute it to "superficial causes which were easily manageable, such as greedy speculators, enemies of the state or fluctuations in the exchange rate".[142] On the other hand "the less production is hampered, the less waste on the part of governments and private individuals is encouraged, the less savings are penalised[143]and the fewer obstacles are placed on the free circulation of goods, capital and labour, the less time the crisis would continue".[144] However, again in the spring of 1921,

Pareto observed that "the steps which continue to be taken do more harm than good, because they compound the problems of production",[145] even if, in the light of the separate need to preserve social equilibrium, "possibly what could be done was done".[146]

In any case, from June 1921 Pareto considered that the various cases of increased working hours and reduced salaries seen on the international stage foreshadowed a move back towards economic equilibrium and an end to the crisis.[147] In April 1922, Pareto observed other signs of a return to equilibrium, including a reduction in private consumption together with attempts to reduce public expenditure, which, however, were fruitless in the absence of "the fortitude to scale back the state, municipal and provincial bureaucracies".[148]

A large part of the Italian economic crisis of the time was provoked by the crisis of the industrial and banking conglomerate Ansaldo-Banca Italiana di Sconto.[149] In December 1921, Pareto, hoping in general terms for the abolition of the prohibition on fixed-term contracts as being "harmful, as well as ineffective",[150] declared that the moratorium for the Banca Italiana di Sconto,[151] while probably "indispensable",[152] was also damaging. Indeed, on the one hand, "abrupt and arbitrary steps like this are highly damaging for credit, both domestically and abroad"[153] and, on the other, by paving the way for a "rescue", they cause "a destruction of wealth which can greatly exceed the little which is preserved".[154]

Soon afterwards, referring to the related collapses of Ansaldo and the Banca Italiana di Sconto, Pareto observed that when "the banks are too close to the government, they are often dragged to their doom".[155] In the case under consideration, since the beginning of the war the Italian government had encouraged industry (as was necessary, for that matter, to try to win the war[156]) "to construct ambitious production facilities which could never prosper in normal times, financed by the banks".[157] The government itself, however, was later to obstruct, "very vigorously, the functioning of the mechanisms which would gradually have returned industry and the banks to a state of normality", specifically by scaling down the former now that the requirements of wartime production no longer applied.[158] In Pareto's view, complicit with the management of the Banca Italiana di Sconto were both the shareholders who had not exercised adequate control and the government which had encouraged it to continue financing Ansaldo.[159] In his opinion, in general terms, this also was due "to a large extent to the government's lack of authority which, in turn, derives from the scarcity of idealism in the country".[160]

The Problems of Public Finance

The post-war disequilibrium in the real economy was caused by the continuing disequilibrium in public finances[161] which, already at the time of the conflict, had led Pareto to predict that all the formerly warring nations, unable to meet the interest payments in hard currency on the enormous public debt[162] which had accumulated during the war, would have to make use of "dubious methods"[163] in order to at least keep up the appearance of meeting their commitments. Pareto nevertheless excluded the option of bankruptcy, since "modern states need credit and thus cannot afford to brutally alienate it".[164]

In general, Pareto advanced the following hypotheses:

1. "The state uses hard currency to pay the interest, leading to increases in taxation" which would be painless for society only if it corresponded with a period of post-war economic prosperity which, however, was difficult to envisage;[165]
2. "To pay off the debt, the state expropriates private capital", which Pareto considered to be "a step which would be very difficult to implement in practice".[166]
3. Lastly, "the state reduces its indebtedness through devaluation of the currency,[167] a solution which, viewed over the course of centuries, could be considered as absolutely normal",[168] since it "has the great advantage of being practically imperceptible for the bulk of the population".[169] This solution would penalise savers who had invested in fixed-rate securities, to be repaid in the devalued currency.[170] However, in Pareto's opinion, "what will be the consequences of this course of action is a question which only the integration of sociology can presume to address".[171]

Basically, in a scenario where the lira of October 1918 was worth half of its value in 1914, a debt incurred in October 1918 at 5% interest "would in fact be incurring 10% interest and thus could potentially be impossible to repay" if the lira were to return to its 1914 value, while if it kept its present value, "the ability or inability to pay the interest on the debt would depend on factors extraneous to the value of the currency" and "if in the future the lira were to diminish in value, it would be easy to pay the interest".[172] Hence, in January 1920, Pareto felt able to underline that "those who lent money to the Austrian and German governments when the crown and the mark were at par with gold, are now, properly speaking, victims of a bankruptcy" on the basis that "the crown has fallen by 5 gold-centimes and the mark by 10 gold-centimes".[173] Similarly, in February 1920, with 100 Italian lire being worth 33.47 Swiss francs[174] (a currency then much nearer to parity with gold than the Italian lira), Pareto pointed out that the interest on Italian public debt, nominally equivalent to 4.5 billion lire, corresponded to only 1.5 billion Swiss francs,[175] whereas it would have been worth 4.5 billion Swiss francs if the lira had the same value as the Swiss franc, that is if 100 lire were worth 100 Swiss francs, as was the case before the war.[176]

As mentioned above, the attainment of macroeconomic equilibrium also requires the elimination of the public deficit, which in Pareto's view could be achieved only through a drastic reduction in public expenditure, even if in June 1922 he was still wondering "whether all this is possible under present political and social conditions, while it is not apparent how these could be readily modified".[177]

Instead, the most significant fiscal measures of the post-war period[178] fall within the "more general problem of determining to what extent tax increases could be beneficial".[179] Thus, in relation to the obligation of a declaration of identity for bond-holders, in Pareto's opinion this would encourage the flight of capital from Italy, effective possible counter-measures to which would also "discourage the arrival of capital from abroad, for which the need has rather increased than diminished".[180] It would hamper the influx of capital "to industrial and commercial concerns, or else it would be necessary to pay more interest in order to obtain the same amount".[181] As regards the law intended to castigate loan sharks,[182] on the other hand, Pareto pointed out that a post-war income of 300,000 lire was equivalent to a pre-war income of 100,000 lire, with the implication that if the rentier in question were deprived of 200,000 at the current value, he would be left with only 33,000 lire at pre-war values, so that a levy of this kind would deprive him "not so much of profits made during the war as of much of the original sum".[183] Instead, as far as public expenditure was concerned, Pareto, after praising Giolitti's government for having reduced the cost of imposing price controls on bread,[184] observed that while this measure had reduced the state's deficit from 14 billion lire to 4 billion lire, it would have completely different repercussions according to whether it was used to stimulate consumption or production.[185]

In the summer of 1922, having estimated the state deficit at 1,500 million gold-francs, Pareto concluded that "it is not in fact that much, so Italy is not so badly-off in comparison to other countries", even if this figure was supposed to be further reduced through additional measures to streamline the bureaucracy, to forego certain planned items of expenditure, to review the remainder and to eliminate the deficits in the public services,[186] although the implementation of these would also pose problems.[187]

Thus, at the beginning of November 1922, in one of his first comments on the new government's economic policy, Pareto said that "the government's scheme for regulating the railway companies and others is excellent, but everything now depends on how it is implemented", reiterating that "the idea of putting the railways in the hands of a railway workers' union should at least be examined".[188] However, soon afterwards he acknowledged that since everyone wants to reduce public expenditure but no one wants their own interests to be affected, "if Mussolini manages to [reduce public expenditure]he will have achieved something in comparison to which the labours of Hercules were insignificant".[189] Thus it was with evident satisfaction that he was able to declare, as early as mid-November, that it appeared that "demagogic finance is coming to an end"[190] and that, dispelling his earlier fears,[191] "Mussolini has made good choices in some cases. For example, De Stefani,[192] Tangorra[193] and others seem to be acting commendably".[194]

In January 1923, Pareto expressed his approval of the government's first tax measures[195] (particularly the abolition of the requirement for identification of bond-holders) because, by ending the oppression of savers in order to finance consumption on the part of the majority, they had saved the country from arriving "at a situation like the one Russia has been brought to by communism".[196] In general terms, where recent governments had appeased demagogic feelings and vested interests, the new government "seeks at least to re-establish a balance among these forces so as not to compromise the nation's prosperity".[197] Further underlining the illusory nature of the fiscal parity which the new government sought to achieve through uniform tax increases, Pareto proposed that moderation should be shown towards agricultural producers (as being fundamental for social stability), salaries (to avoid increases in costs of production and hence in prices, which would affect "economic prosperity") and capital (to avoid a "deluge of economic and financial problems" in a nation which was short of it).[198]

In the following June, commenting on the budget forecast for the 1923–1924 financial year which had just been released by De Stefani, Pareto noted that while the accounting deficit "is probably not too far from the official figure of 2,616 million lire, there still remains the critical problem of the correspondence between the accounts and reality".[199]

In this regard, Pareto doubted, first of all, whether it was realistic to include among exceptional revenues the billion lire which it was hoped could be obtained in war reparations.[200]

Furthermore, and more particularly, having noted that the relationship between the values for paper money (as used in the budget) and for gold was highly variable, he pointed out that, consequently, the real value of the fixed amounts[201] in the budget entries must likewise be highly variable and that therefore it was necessary to "be very cautious in interpreting the data from the budget".[202] Accordingly, on the basis, for example, that on 6 June 1923 an Italian lira was quoted at 0.24 Swiss gold-francs on the Geneva stock exchange, it followed that the effective deficit was between 2,616 million lire and 654 million Swiss gold-francs, which would confirm "that in the end a deficit of this order [is] not in fact so large nor such as to exceed the country's resources".[203] On the other hand, if the paper-lira were to regain parity with the gold-lira, Italian public debt (which at that time stood at 5,500 paper-lire and thus at 1,375 million Swiss gold-francs) would be quadrupled, leading to the "bankruptcy of the Italian state".[204]

Having said this, as an application of the principle of equality, Pareto praised De Stefani's introduction of new taxes both on independent farmers (although this was fairly insignificant in view of its moderation) and on the salaries of public employees (this measure meeting with Pareto's approval as contributing, "albeit only negligibly", to the reduction of consumption).[205] In general, insofar as De Stefani "is aiming to restore the national finances and economy to health and is doing so efficaciously, the measures he is introducing are constructive".[206] In terms of political economy, it was "to Fascism's great credit that it had mediated between objective and subjective views".[207]

Devaluation and Inflation

In April 1919, Pareto stated that following gold's loss of its status as an international currency due to the monopoly of use the various nations had adopted from the time of the war, the determination of a currency's value could be performed only by comparing it with that of the currency whose value varied least in relation to gold, that is the Swiss franc.[208] On this basis, Pareto pointed out that on the Geneva stock exchange, the price of 100 Italian lire in Swiss francs had risen from 43.70 on 20 July 1918 to 77.75 on 14 October 1918, before stabilising at 75.20 (12 March 1919) and then dropping back to 64.55 (2 April 1919).[209] This increase in the value of the lira was due to Italy's victory in the conflict, but also to the Italian government's defence of the currency, in which, incidentally, it had acted like all the other governments of the Entente, with the reduction in this support causing the later devaluation.[210] In any case, Pareto was of the view that a lasting appreciation of the currency was achievable "only by increasing production or by reducing consumption", whereas in Italy production had been hampered by legislation which had favoured labour over capital and property during the early post-war years.[211] Thus, Pareto observed in September 1920 that on the Geneva stock exchange, the price of 100 lire in Swiss francs had fallen from 74.75 to 20.20 in the period from 2 January 1919 to 1 September 1920, notwithstanding all the government's efforts to avoid this kind of devaluation, including the issue of loans (to soak up paper money), tax increases (to reduce the public deficit and the resulting need to finance it through issues of paper money) and exchange controls.[212]

Pareto thought that the general devaluation of currencies in relation to their respective pre-war gold parities, which was affecting all countries with the exception of the United States, "indicates a reduction in economic prosperity",[213] resulting from the destruction of wealth caused by the war and by the colossal, ill-advised public expenditure of the post-war period.[214] Thus, in April 1922, in reference to the proposals made at the Genoa Conference for the joint reintroduction of stable parities with gold, Pareto commented that what was needed above all was specific and credible details of how to restore state finances to parity, how to re-establish the required foreign currency reserves and how to use these to defend the new rates.[215] However, despite his generally highly favourable attitude towards the debut of Mussolini's first government, Pareto foresaw that "the threatened measures for the control of the exchange rate" would unintentionally further devalue the lira.[216]

Repeating that "the depreciation of the currency corresponds to increases in certain prices",[217] Pareto was hence of the view that high costs of living[218] were accounted for by everything (specifically ill-conceived government measures) which interfered with production and which encouraged "extravagant" consumption.[219] Accordingly, Pareto asserted that, for example, "any measures which reduced the purchasing power of the nouveaux riches and of the well-off working class would be more efficacious than the most ingenious price control schemes that could be dreamt up".[220]

However, in June 1920 Pareto pointed out that although the ongoing fall in prices was "beneficial for a large number of people", it also had negative aspects because, in real terms, "producers, labour, capitalists and taxpayers (i.e. the "most dynamic and hence dominant section of the population")" were compromised by this phenomenon, while only savers, the least dynamic category, were rewarded.[221] As a result, "governments face increasing difficulties in seeking to restrain the dominant part of the population with the help of the weakest".[222]

Notes

1. For guidance for the study of this period see the classic (Maier 1975).
2. Pareto to Placci, 7th November 1918, see (Pareto 1975, p. 1007); see (Pareto 1920a, reprinted in Pareto 1987, p. 995).
3. Pareto to Bodio, 20th February 1920, see (Pareto 2001, p. 347).
4. See (Pareto 1920b, reprinted in Pareto 1980, pp. 931–932).
5. Ibid., pp. 937–938.
6. Ibid., p. 938.
7. Ibid., p. 939.
8. Ibid., pp. 953–954.
9. Ibid., p. 940.
10. See (Pareto 1919, reprinted in Pareto 1920c, p. 237).
11. Ibid., p. 240. In Pareto's view "America imposes its will through the concession or denial of credit and through export permits and bans" while England exploits its position of strength, "in the absence of the formidable and fearsome competition from Germany", to "force the other nations to swell the earnings of the alliance between its plutocrats and its workers", Ibid., p. 237.
12. Ibid., p. 240. Pareto emphasised that "there is no historical basis for the truth of the notion that democracies are less warlike than monarchies", see (Pareto 1921a, reprinted in Pareto 1987, p. 692).
13. See (Pareto 1919a, reprinted in Pareto 1920c, p. 242).
14. See (Pareto 1920d, reprinted in Pareto 1987, p. 572). In two historic votes held on 19th November 1919 and 19th March 1920, United States Senate, with its Republican majority, had refused to approve the country's entry into the League of Nations.

15. Indeed, in Pareto's view (see Pareto 1918a, reprinted in Pareto 1920c, p. 172), the principle in question, based on the idea that "every nation should be sovereign and independent", was undermined by the vagueness of the term nation, due to the increased ambiguity of terms such as race, religion, language and historical tradition".
16. Ibid., continuation of note 1 on p. 371.
17. Ibid.
18. See (Pareto 1920e, reprinted in Pareto 1966, p. 313), (Pareto 1922a, reprinted in Pareto 1966, p. 329).
19. On the other hand, in September 1922, Pareto would advance Turkey's military revolt against "the treaty of Sèvres, which reduced it to slavery" among the examples of "the effectiveness of the use of force for the realignment of the political and social situation", see (Pareto 1922b, reprinted in Pareto 1987, p. 773).
20. See (Pareto 1922c, reprinted in Pareto 1987, p. 701).
21. Ibid., pp. 701–703.
22. To be held from 10 to 19 May 1922.
23. See (Pareto 1922d, reprinted in Pareto 1987, p. 726).
24. Ibid.
25. Ibid., pp. 726–727.
26. Ibid., p. 728.
27. See (Pareto 1922e, reprinted in Pareto 1987, p. 728).
28. See (Pareto 1923a, reprinted in Pareto 1987, pp. 780–781); Pareto to Vittore Pansini, 12th October 1919, see (Pareto 1989, p. 678).
29. See (Pareto 1922e, reprinted in Pareto 1987, p. 732).
30. The peace treaty of Versailles which was later imposed on the Germans was signed by the victors on 28th June 1919. For a useful bibliographic overview of the topic see (Boemeke et al. 1998).
31. See (Pareto 1922f, reprinted in Pareto 1987, p. 745).
32. Formalised in the treaty of Rapallo of 16th April 1922.
33. Ibid., and see (Pareto 1921a, reprinted in Pareto 1987, p. 692).
34. Pareto to Naville, 9th May 1923, see (Pareto 1975, p. 1145). Thus, his immediate resignation (tendered on 8th February 1923, two days after his nomination had been finalised) from the League of Nations mixed preparatory commission on arms limitation, which he had agreed to join at the personal request of Mussolini (Pareto to Mussolini, 28th December 1922, ibid., pp. 1123), while mainly due to health reasons, Pareto to Eric Drummond, 8th February and 15th March 1923, ibid., pp. 1131–1132, 1138 was also partly due to his impression that participating in a commission thus constituted would be "a waste of time", Pareto to Naville, 9th May 1923, ibid., p. 1145.
35. See (Pareto 1920f, reprinted in Pareto 1980, p. 926).

36. See (Pareto 1920d, reprinted in Pareto 1987, pp. 574–575). For Keynes' well-known, and comparable, interpretation, see (Keynes 1920). For an innovative reformulation of the question, see (Ritschl 1999).
37. See (Pareto 1922d, reprinted in Pareto 1987, p. 728).
38. See (Pareto 1923a, reprinted in Pareto 1987, p. 782). Pareto further pointed out that if the victorious nations were to accept German exports in payment for war debts associated with the conflict it would cause serious difficulties for industries in those countries. If, on the other hand, German exports were not accepted, the victorious nations would face not only financial difficulties but also economic, in that they would not be able to export to Germany, which would not possess the currency needed to pay for its imports, see (Pareto 1921b, reprinted in Pareto 1987, p. 670).
39. See (Pareto 1922g, reprinted in Pareto 1987, p. 771).
40. See (Pareto 1923b, reprinted in Pareto 1987, p. 788). It is remembered that the French and Belgian armies occupied the Ruhr mining region on 11th January 1923 in retaliation for the suspension in war reparation payments decided unilaterally by the German government.
41. See (Pareto 1922d, reprinted in Pareto 1987, p. 728); (Pareto 1923b, reprinted in Pareto 1987, p. 789); (Pareto 1922h, reprinted in Pareto 1987, p. 761).
42. Pareto to Pansini, 11th June 1917, see (Pareto 1989, p. 579). For a valuable introduction to the latest literature on the October Revolution, see (Cigliano 2018).
43. Pareto to Pansini, 25th June 1917, see (Pareto 1989, p. 602).
44. Pareto to Pansini, 23rd November 1917, ibid., p. 626.
45. In February 1917, after speculating that "if the Tsarist regime had not gone to war, it might still be in power", Pareto in any case felt "more doubt than conviction" that it was "Russia's destiny that [the Tsar] entered the war and was overthrown", Pareto to Pansini, 9th December 1917, ibid., p. 630.
46. In October 1917.
47. In September 1917.
48. See (Pareto 1920c, p. 351). Further, if Lenin were to fall "it will have been the military rather than the intellectuals who have destroyed his government", leaving open "the question of whether this was for the better or the worse", see (Pareto 1920g, reprinted in Pareto 1966, p. 308).
49. See (Pareto 1922b, reprinted in Pareto 1987, p. 773).
50. See (Pareto 1919a, p. 243).
51. Ibid., p. 247.
52. Ibid.
53. Ibid., p. 249.

54. Ibid.
55. Ibid., p. 249.
56. See (Pareto 1919b, reprinted in Pareto 1920c, pp. 259–260).
57. On 10th April 1918.
58. See (Pareto 1918b, reprinted in Pareto 1920c, p. 170).
59. See (Pareto 1918a, reprinted in Pareto 1920c, pp. 182–183). Pareto considered the Bolshevik revolution camparable "to the early French revolution", see (Pareto 1919b, reprinted in Pareto 1920c, p. 257).
60. Pareto to Pippo Naldi, 17th January 1919, BPS-la.
61. See (Pareto 1922d, reprinted in Pareto 1987, p. 728).
62. Representing for the West "the greatest danger for the future, although overlooked", see (Pareto 1923, reprinted in Pareto 1987, p. 786).
63. Ibid., pp. 785–786.
64. See (Pareto 1921c, reprinted in Pareto 1987, p. 1059).
65. On the other hand, in March 1923, in relation to a possible resumption of commercial relations between the Western countries and Russia, Pareto said that "the Russian market is poor and will absorb and supply very little overall", see (Pareto 1923c, reprinted in Pareto 1987, p. 785).
66. See (Pareto 1922i, reprinted in Pareto 1987, p. 754).
67. See (Pareto 1923d, reprinted in Pareto 1987, p. 776).
68. On this subject see the classic (Spriano 1964).
69. See (Pareto 1920h, reprinted in Pareto 1980, p. 1011).
70. See (Pareto 1920i, reprinted in Pareto 1980, p. 1020).
71. Ibid.
72. Ibid., p. 1021.
73. See (Pareto 1920l, reprinted in Pareto 1980, p. 1028).
74. See (Pareto 1920m, reprinted in Pareto 1980, p. 1031).
75. Ibid., p. 1032.
76. Ibid., p. 1033.
77. Which duly occurred in April 1921, see (Pareto 1921b, reprinted in Pareto 1987, p. 671).
78. See (Pareto 1920m, reprinted in Pareto 1980, pp. 1033–1034).
79. See (Pareto 1921d, reprinted in Pareto 2005, p. 271).
80. Ibid.
81. Ibid., p. 273.
82. Ibid., p. 274.
83. Pareto claimed to "view Fascism from the same absolutely objective perspective he had adopted in his scrutiny of multifarious other political, economic and social phenomena", see (Pareto 1923e, reprinted in Pareto 1987, p. 738). A substantial literature is extant on Pareto's attitude towards Fascism: for some interesting examples, see (Montini 1974) and (Barbieri 2003).

84. Pareto to Linaker, 18th March 1921, see (Pareto 1975, p. 1062).
85. Pareto to Linaker, 19th April 1921, ibid., p. 1063.
86. Pareto to Linaker, 6th July 1921, ibid., pp. 1068–1069; see (Pareto 1922l, reprinted in Pareto 1980, pp. 1079, 1084–1085); Pareto to Pantaleoni, 11th August 1922, see (Pareto 1984, p. 309); see (Pareto 1922a, reprinted in Pareto 1966, p. 324).
87. See (Pareto 1921e, reprinted in Pareto 1980, p. 1051).
88. In what was to be his last government, from 15th June 1920 to 4th July 1921.
89. Ibid.
90. Ibid., p. 1052.
91. See (Pareto 1922l, reprinted in Pareto 1980, p. 1088). Previously, Pareto had also admired the foresight with which the plutocrats had aided Mussolini, following his expulsion from the Italian Socialist Party, to found his newspaper "Il Popolo d'Italia" (whose first issue appeared on 15th November 1914), demonstrating once again how they were able to "get their mitts in everywhere", Pareto to Pantaleoni, 30th August 1921, see (Pareto 1984, p. 292).
92. Pareto to Tommaso Giacalone-Monaco, 1st June 1922, see (Pareto 1989, p. 754).
93. Pareto to Pantaleoni, 17th August 1922, see (Pareto 1984, p. 311); Pareto to Vincenzo Fani, 11th October 1922, see (Pareto 1975, p. 1106).
94. See (Pareto 1922m, reprinted in Pareto 1987, p. 768).
95. Pareto to Linaker, 11th September 1922, see (Pareto 1975, p. 1097).
96. A specialist in the history of ancient Greece and at that time a nationalist-leaning member of parliament, Ciccotti had long been of the socialist persuasion and had been Pareto's guest in Lausanne in the summer of 1898, when he was wanted by the Italian authorities for his part in the civil disorders of the time in Milan.
97. In the article Audacious letter from the honourable E. Ciccotti. A year of dictatorship to rescue Italy. The nation will respond to the summons (Ardita lettera dell'on.E.Ciccotti. Un anno di dittatura per salvare l'Italia. Il Paese, interpellato, accoglierà la proposta). *Il Giornale d'Italia*, 22nd June 1922.
98. See (Pareto 1922b, reprinted in Pareto 1987, p. 773).
99. Pareto to Fani, 11th October 1922, see (Pareto 1975, p. 1106).
100. Ibid.
101. Pareto to Pantaleoni, 17th October 1922, see (Pareto 1984, p. 313).
102. Ibid.
103. Ibid., p. 315.
104. Ibid., p. 316.

105. See (Pareto 1923f, reprinted in Pareto 1980, p. 1155). Pareto was probably alluding to the article *The sociological idea of the state* (*Il concetto sociologico dello Stato)* which Fani, under the pseudonym of Volt, had published on pp. 422–426 of the 1922 edition of "Gerarchia".
106. See (Pareto 1923g, reprinted in Pareto 1974, p. 190).
107. Ibid.
108. On 31st October 1922.
109. Pareto to Ettore Rosbock, 16th November 1922, see (Pareto 1989, p. 763).
110. Pareto to Lello Gangemi, 13th November 1922, ibid., p. 1114. Hence Pareto expressed the wish to "help, not harm, the best government that Italy has had for many years", Pareto to Scalfati, 24th November, 2nd December 1922, see (Pareto 2001, pp. 396–397). He thus considered his proposed appointment as Senator inopportune, especially with his recent adoption of the citizenship of Fiume in order to obtain a divorce from his first wife in that city, where the unprecedented legislation introduced under D'Annunzio's administration of Carnaro was still in force. On 1st March 1923 the nomination was nevertheless approved but was not ratified due to the non-arrival of the required documents, since Pareto had become a citizen of Fiume. In order to obtain "the great satisfaction of making [his own small] contribution to the great edifice rising up in Italy", Pareto to Scalfati, 2nd December 1922, ibid., p. 397, Pareto declared his willingness, on a strictly unremunerated basis, to be nominated an "specialist consultant on economic and financial matters with the Italian legations either in Bern or with the United Nations in Geneva", ibid. and note 35 above.
111. Pareto to Pantaleoni, 23rd December 1922, see (Pareto 1984, p. 320).
112. See (Pareto 1923g, reprinted in Pareto 1974, p. 192).
113. See (Pareto 1923f, reprinted in Pareto 1980, p. 1159). According to Pareto, by June 1923 there was only the Fascist party apart from the popular party and he acknowledged Sturzo's skill in setting it up, managing it "prudently" in difficult circumstances and keeping it together despite its numerous splinter groups, see (Pareto 1923h, reprinted in Pareto 1966, p. 193). Again, in Pareto's view, the two parties disagreed mainly over the electoral law, with the popular party wishing to maintain the proportional system (since it owed its importance to this law and saw its future prospects as being tied to it), while the Fascist party hoped for additional seats in parliament for the party holding a relative majority, convinced that such a system would allow it to consolidate its grip on power definitively, ibid., p. 194. Nevertheless, were the powers of parliament to be curtailed (which, with the current setup, might yet cause the

overthrow of the new government), then the electoral question would become of secondary importance, ibid.

114. See (Pareto 1923f, reprinted in Pareto 1980, p. 1159).
115. As shown by the case of the Italian electorate which, democratic prior to the war, became socialist in the period 1919–1920 on the basis of the unfulfilled promises of greater well-being received during the war, before finally gravitating to Fascism following similar disappointments with socialism (Pareto 1923h, reprinted in Pareto 1966, p. 195).
116. Ibid., p. 193.
117. See (Pareto 1923i, reprinted in Pareto 1987, p. 798).
118. Ibid., p. 796.
119. Ibid., p. 797. Similarly, the powers of the town councils would need to be drastically curtailed, ibid., p. 799.
120. Ibid., p. 797.
121. At the time its members were appointed by the king acting on government proposals.
122. Ibid., p. 797.
123. Ibid., p. 800.
124. See (Pareto 1923l, reprinted in Pareto 2005, 287).
125. Ibid., p. 288.
126. Ibid. On this topic in general see (Cassese and Dente 1971).
127. See (Pareto 1922n, reprinted in Pareto 1987, p. 709). Similarly, immediately after the conflict, see (Pareto 1919a, reprinted in Pareto 1920c, p. 230); Pareto to Linaker, 16th August 1920, see (Pareto 1975, p. 1043).
128. See (Pareto 1919a, reprinted in Pareto 1920c, p. 231)
129. Ibid., p. 241.
130. See (Pareto 1922o, reprinted in Pareto 1980, p. 1111); (Pareto 1923m, reprinted in Pareto 1974, pp. 197–198).
131. See (Pareto 1922o, reprinted in Pareto 1980, p. 1112).
132. Pareto took the opportunity to repeat that "ideas for restoring production fluctuate between two extremes: on the one hand, leaving the mechanism of demand and prices to operate freely or, alternatively, leaving the regulation of labour and capital to the government, as Lenin does", see (Pareto 1920n, reprinted in Pareto 1980, pp. 607–608).
133. Ibid.
134. See (Pareto 1920o, reprinted in Pareto 1987, p. 570).
135. See (Pareto 1920p, reprinted in Pareto 1987, p. 592).
136. See (Pareto 1920q, reprinted in Pareto 1987, pp. 600–601).
137. See (Pareto 1921f, reprinted in Pareto 1987, p. 668.
138. See (Pareto 1921g, reprinted in Pareto 1987, p. 693); (Pareto 1921h, reprinted in Pareto 1987, p. 673).

139. See (Pareto 1920r, reprinted in Pareto 1987, p. 647).
140. Ibid., p. 648.
141. Ibid.
142. See (Pareto 1920s, reprinted in Pareto 2005, p. 257).
143. This affected him directly because the Swiss socialist party had gathered the requisite number of signatures to submit a popular party-inspired draft bill proposing "the imposition of a tax on wealth" to a referendum. In particular, this proposal provided for a special progressive tax on the capital of physical or legal persons, varying from 8% to 60%. To avoid the risk of having to incur this tax, Pareto decided at first to transfer his place of residence to Thonon-les-Bains (on the French shores of Lake Leman, opposite Céligny), later abandoning this destination in favour of the much-nearer trans-frontier area of Divonne-les-Bains. He planned to return to Switzerland if the proposal was rejected but to remain permanently in France if it was approved, Pareto to Pantaleoni, 17th October 1922, see (Pareto 1984, p. 312); Pareto to George-Henri Bousquet, 20th August 1922, see (Pareto 1975, p. 1107); Pareto to Guido Sensini, 18th November 1922 (see Pareto 1975, p. 1115). On 3rd December 1922, the proposal was rejected by 736,952 electors and approved by only 109,702 and was rejected by all the cantons and semi-cantons.
144. See (Pareto 1920s, reprinted in Pareto 2005, p. 257).
145. See (Pareto 1921f, reprinted in Pareto 1987, p. 665); (Pareto 1922d, reprinted in Pareto 1987, p. 727); (Pareto 1921d, reprinted in Pareto 2005, p. 274).
146. See (Pareto 1921f, reprinted in Pareto 1987, p. 667.
147. See (Pareto 1921b, reprinted in Pareto 1987, p. 669); (Pareto 1922p, reprinted in Pareto 2005, p. 278).
148. See (Pareto 1922q, reprinted in Pareto 1987, p. 733).
149. On this episode see (Doria 1989) and (Falchero 1990).
150. See (Pareto 1921i, reprinted in Pareto 1987, p. 699).
151. The moratorium was not approved by Ivanoe Bonomi's government, leading to the bankruptcy of the Milan bank.
152. Ibid.
153. Ibid.
154. Ibid.
155. See (Pareto 1922n, reprinted in Pareto 1987, p. 707).
156. Ibid., p. 710.
157. Ibid., p. 708.
158. Ibid., pp. 708–709.
159. Ibid., p. 711.
160. Ibid., p. 712.

161. Accordingly, if Y = production, C = private consumption, G = public expenditure and T = tax revenues, with G–T thus corresponding to public consumption, a macroeconomic deficit will ensue from the occurrence of the disparity $C + (G - T) > Y$.
162. According to Pareto "loans are a way to induce populations to accept what they would not accept in the form of taxation" and, more specifically, given the unlikelihood of the state repaying the capital received and the lengthy timescale for repayments of interest, "a means of taking money from one part of the population, without causing too many protests, to the benefit of another part of the population", which would do the nation harm or good "according to the extent to which the classes rendering the maximum of utility are alienated or accommodated", Pareto to Benvenuto Griziotti, 5th October 1917, see (Pareto 1975, p. 989). More generally, Pareto observed that any measure in political economy will have not only economic but also social consequences (i.e. relating to feelings as well as private interests) and that often a measure cannot maximise both these types of effects simultaneously. This common "discrepancy is [normally] to the detriment of the former category [but it] can serve to purchase the social harmony which would be jeopardised if feelings and interests were impinged on", see (Pareto 1920t, reprinted in Pareto 1987, p. 628).
163. Pareto to Walter Eggenschwyler, 10th March 1915, see (Pareto 1989, p. 551).
164. See (Pareto 1916, reprinted in Pareto 1920c, p. 59). More generally, Pareto noted that in the nineteenth century the governments of civilised countries had been better able to meet their commitments by issuing bonds than by printing paper money, see (Pareto 1919c, reprinted in Pareto 1920c, p. 191).
165. See (Pareto 1918c, reprinted in Pareto 1920c, p. 209).
166. Ibid., p. 210.
167. Accordingly, the devaluation of a currency leads to a corresponding reduction in the purchasing power of the interest on public debt as well as in the principal itself, see (Pareto 1916, reprinted in Pareto 1920c, p. 60).
168. See (Pareto 1918c, reprinted in Pareto 1920c, p. 212).
169. Ibid., p. 213.
170. See (Pareto 1916, reprinted in Pareto 1920c, p. 61).
171. Ibid., p. 62.
172. See (Pareto 1918c, reprinted in Pareto 1920c, p. 205).
173. See (Pareto 1921f, reprinted in Pareto 1987, p. 837).
174. See (Pareto 1920u, reprinted in Pareto 1987, p. 576).
175. Ibid., p. 577, because $(4{,}500{,}000{,}000 \times 33.47)/100 = 1{,}506{,}150{,}000$.

176. Ibid., p. 578.
177. See (Pareto 1922r, reprinted in Pareto 1987, p. 757).
178. In regard to which, see (Marongiu 2019, pp. 261–267).
179. See (Pareto 1920v, reprinted in Pareto 1987, p. 641).
180. Ibid.
181. Ibid., p. 642.
182. That is the war profiteers, in alliance with the plutocrats, who, in Pareto's view, were hated mainly because "they were partly responsible for the economic changes from which they benefitted", see (Pareto 1920z, reprinted in Pareto 1987, p. 643).
183. Ibid., p. 646. In any case, as early as April 1921 Pareto noted that enthusiasm for the confiscation of war profits was waning due to the difficulties of implementing the measure. See (Pareto 1921b, reprinted in Pareto 1987, p. 671).
184. See (Pareto 1921l, reprinted in Pareto 1987, p. 661).
185. Ibid., pp. 661–662.
186. See (Pareto 1922s, reprinted in Pareto 1987, pp. 759–760).
187. In fact, "the many forces acting to encourage expenditure increase with the diminishing authority of the state", see (Pareto 1922t, reprinted in Pareto 1987, p. 764).
188. Pareto to Giuseppe Stanislao Scalfati, 7th November 1922, see (Pareto 2001, p. 393). It should be remembered that the government preferred to reorganise the sector with the sacking of 36,000 railway workers in January 1923.
189. Pareto to Pantaleoni, 30th October 1922, see (Pareto 1984, p. 317).
190. Pareto to Pietri-Tonelli, 13th November 1922, see (Pareto 1975, p. 1113).
191. Pareto to Scalfati, 1st November 1922, see (Pareto 2001, p. 393).
192. An early follower of Fascism, the Padua's university professor in financial science Alberto De Stefani was Minister of Finance and of the Treasury until July 1925. On De Stefani's financial policies, see (Marcoaldi 1980) and (Fausto 2007).
193. An economist and a member of the popular party, Vincenzo Tangorra held the position of Minister of the Treasury until his sudden death in December 1922.
194. Pareto to Scalfati, 21st November 1922, see (Pareto 2001, p. 395).
195. According to Pareto the new cabinet was able to decide these because it was "fortified by faith and ready to use force", see (Pareto 1923n, reprinted in Pareto 1987, p. 1153).
196. Ibid., pp. 1150–1151.
197. Ibid., p. 1152.
198. Ibid., p. 1153.

199. See (Pareto 1923o, reprinted in Pareto 1980, p. 1186).
200. Ibid.
201. Ibid., p. 1187.
202. Ibid.
203. Ibid., p. 1188.
204. Ibid.
205. Ibid., p. 1189.
206. Ibid., p. 1190.
207. Ibid.
208. See (Pareto 1919d, reprinted in Pareto 1987, p. 217).
209. Ibid.
210. Ibid., p. 218. This support for the lira was in turn implicitly motivated by the "purpose of sustaining popular morale during the war", ibid. However, at the beginning of May 1918, Pareto had noted that the National Institute for Foreign Exchange had begun its operations of support for the lira in mid-April when 100 lire had been worth 51 Swiss francs in Geneva. This rate had already decreased to 46.55 by May, Pareto to Bodio, 2nd May 1918, see (Pareto 2001, p. 314), Pareto to Pantaleoni, 2nd May 1918, see (Pareto 1984, p. 231).
211. See (Pareto 1920aa, reprinted in Pareto 1987, p. 651).
212. Ibid., p. 650.
213. See (Pareto 1921m, reprinted in Pareto 1987, p. 688).
214. Ibid., p. 687.
215. See (Pareto 1922u reprinted in Pareto 2005, p. 279).
216. Pareto to Pantaleoni, 30th October 1922, see (Pareto 1984, p. 316).
217. See (Pareto 1921n, reprinted in Pareto 2005, p. 274).
218. Pareto stipulated that "it is impossible to identify a single index for prices. There are many, each serving for specific purposes", see (Pareto 1919e, p. 102). On this basis, he deemed admissible the statistician Costantino Ottolenghi's recent idea of constructing an index which, on the basis of the importance of the role the various commodities "play in foreign and domestic trade", ibid., p. 102, sought neither to "completely exclude nor to give excessive prominence to the individual quantities of commodities under consideration", ibid., p. 104. Pareto added that the doubling of real prices (i.e. when expressed in Swiss francs, the currency which had lost least value in comparison to gold) in the various countries between 1913 and 1919 also represented an indicator of the damage wrought on the civilised nations by the war, see (Pareto 1920ab, reprinted in Pareto 1987, pp. 629–631).
219. See (Pareto 1919f, reprinted in Pareto 1987, p. 558).
220. See (Pareto 1919g, reprinted in Pareto 1920c, p. 275). Pareto considered that increases in the money supply contributed only indirectly to

price increases because they permitted the government to cause the destruction of wealth, "which itself was among the immediate causes of price increases", Pareto to Bodio, 12th October 1919, see (Pareto 2001, p. 336).

221. See (Pareto 1920ac, reprinted in Pareto 1987, pp. 636–637).

222. Ibid., p. 638.

Bibliography

Barbieri, Giovanni. 2003. *Pareto e il fascismo [Pareto and Fascism]*. Milano: Franco Angeli.

Boemeke, Manfred F., Gerald D. Feldman, and Elisabeth Glaser. 1998. *The Treaty of Versailles: A Reassessment after 75 Years*. New York and Washington, DC: Cambridge University Press.

Cassese, Sabino, and Bruno Dente. 1971. Una dicussione del primo ventennio del secolo: lo stato sindacale [A discussion of the first twenty years of the century: the Trade Union State]. *Quaderni storici*, 6, September-December, pp. 943–970.

Cigliano, Giovanna. 2018. La rivoluzione russa del 1917 nei recenti orientamenti storiografici internazionali [Recent international historiographical approaches to the Russian revolution of 1917]. *Ricerche di storia politica*, August, pp. 171–190.

Doria, Marco. 1989. *Ansaldo: l'impresa e lo Stato [Ansaldo: the company and the state]*. Milano: Franco Angeli.

Falchero, Anna Maria. 1990. *La Banca Italiana di Sconto, 1914–1921: sette anni di guerra [The Banca Italiana di Sconto, 1914–1921: seven years of war]*. Milano: Franco Angeli.

Fausto, Domenicantonio. 2007. La finanza pubblica fascista [Public finance under Fascism]. In *Intervento pubblico e politica economica fascista [State intervention and Fascist economic policy]*, ed. D. Fausto, 579–585. Milano: Franco Angeli.

Keynes, John Maynard. 1920. *The economic consequences of the peace*. London: Macmillan.

Maier, Charles S. 1975. *Recasting bourgeois Europe: stabilization in France, Germany, and Italy in the decade after world war 1*. Princeton: Princeton University Press.

Marcoaldi, Franco. 1980. Maffeo Pantaleoni, la riforma finanziaria e il governo fascista nel periodo dei pieni poteri, attraverso le lettere ad Alberto De' Stefani [Maffeo Pantaleoni, financial reform and the Fascist government in the period of its absolute power, through his letters to Alberto De' Stefani]. *Annali della Fondazione Luigi Einaudi* 14: 609–666.

Marongiu, Giovannni. 2019. *La politica fiscale nell'Italia liberale e democratica*. Torino: Giappichelli.

Montini, Luigi. 1974. *Vilfredo Pareto e il fascismo [Vilfredo Pareto and Fascism]*. Rome: Volpe.

Pareto, Vilfredo. 1916. I debiti pubblici dopo la guerra [Public debt after the war]. *Rivista di scienza bancaria*, February-March, pp. 93–97.

———. 1918a. Il supposto principio di nazionalità [The supposed principle of nationality]. *Rivista d'Italia*, July 31.

———. 1918b. Après quatre ans de guerre [After four years of war]. *Coenobium*, July-August, pp. 1–24.

———. 1918c. Il futuro delle finanze dello Stato [The future of public finances]. *L'Economista*, October 13, pp. 458–461.

———. 1919a. Speranze e disinganni [Hopes and disillusionment]. *Rivista d'Italia*, marzo 31.

———. 1919b. Il fenomeno del bolscevismo [The phenomenon of Bolshevism]. *Rivista di Milano*, May 20, pp. 71–82.

———. 1919c. *(Preface) Préface* to Bo Gabriel de Montgomery, *Politique financière d'aujourd'hui principalement en considération de la situation financière et économique en Suisse [Contemporary financial policy in the light of the financial and economic situation in Switzerland]*. Paris-Neuchâtel: Attinger, pp. V–XIV.

———. 1919d. Errori sul cambio [Mistakes over the exchange rate]. *Il Tempo*, April 14.

———. 1919e. Considerazioni sul calcolo dell'indice ponderato dei prezzi [Considerations on the calculation of the weighted index for prices]. *Rivista italiana di sociologia*, January, pp. 102–104.

———. 1919f. Contraddizioni [Contradictions]. *Il Resto del Carlino*, August 12.

———. 1919g. Cose vecchie sempre nuove [Old things, new things]. *Il Resto del Carlino*, July 24.

———. 1920a. Dove andiamo? [Where are we going?]. *Il Resto del Carlino*, June 9.

———. 1920b. Trasformazioni della democrazia II. Sgretolamento dell'autorità centrale [Transformations of democracy II. The crumbling of central authority]. *Rivista di Milano*, May 20 and June 5, pp. 45–53, 91–100.

———. 1920c. *Fatti e Teorie [Facts and theories]*. Florence: Vallecchi.

———. 1920d. Utopie [Utopias]. *Il Resto del Carlino*, February 12.

———. 1920e. Une campagne anti-française [An anti-French campaign]. *L'action nationale*, September, pp. 289–295

———. 1920f. Trasformazioni della democrazia I. Generalità [Transformations of democracy I. Overview]. *Rivista di Milano*, May 5, pp. 5–14.

———. 1920g. Respublica literatorum. Réponse à l'enquête [Respublica literatorum. Reaction to the enquiry]. *Les lettres*, January 1, pp. 25–28.

———. 1920h. Il valore di un episodio [The value of an episode]. *Il Resto del Carlino*, September 10.

———. 1920i. I problemi del controllo [The problems of control]. *Il Resto del Carlino*, September 29.

———. 1920l. I problemi sociali del controllo [The social problems of control]. *Il Resto del Carlino*, October 24.
———. 1920m. I problemi economici del controllo [The economic problems of control]. *Il Resto del Carlino*, October 24.
———. 1920n. I rimedi del Manifesto [The remedies of the Manifesto]. *Il Resto del Carlino*, March 25.
———. 1920o. Rimedi alla crisi economica [Remedies for the economic crisis]. *Il Resto del Carlino*, January 28.
———. 1920p. Nuove delusioni [More disappointments]. *Il Resto del Carlino*, March 2.
———. 1920q. E' nato il topolino [A mouse is born]. *Il Resto del Carlino*, March 17.
———. 1920r. Il denaro si fa caro [The rising cost of money]. *Il Resto del Carlino*, July 21.
———. 1920s. La veniente crisi [The coming crisis]. *Il Tempo*, October 17.
———. 1920t. I provvedimenti finanziari. Le restrizioni [The financial measures and the restrictions]. *Il Resto del Carlino*, May 4.
———. 1920u. Bilancio dello Stato [The national budget]. *Il Resto del Carlino*, February 22.
———. 1920v. La nominatività dei titoli [The identification of bond-holders]. *Il Resto del Carlino*, July 2.
———. 1920z. Alto mare finanziario [Financial storm]. *Il Resto del Carlino*, July 4.
———. 1920aa. Deprezzamento della lira [Depreciation of the lira]. *Il Tempo*, September 20.
———. 1920ab. Il futuro dei prezzi [The future of prices]. *Il Resto del Carlino*, May 21.
———. 1920ac. La discesa dei prezzi [The decrease in prices]. *Il Resto del Carlino*, June 23.
———. 1921a. Ugo Stinnes & C.i [Ugo Stinnes and co]. *Il Resto del Carlino*, December 6.
———. 1921b. Miglioramento alla crisi [Relief of the crisis]. *Il Resto del Carlino*, June 2.
———. 1921c. Sgretolamento dello Stato [The crumbling of the State]. *Il Resto del Carlino*, October 4.
———. 1921d. L'avvenire economico e sociale [The economic and social future]. *La critica politica*, February 16, pp. 49–51.
———. 1921e. Due uomini di stato [Two statesmen]. *La Ronda*, July, pp. 437–448.
———. 1921f. Crisi [Crisis]. *Il Resto del Carlino*, April 3.
———. 1921g. Il futuro dell'economia europea [The future of the European economy]. L'*Economista*, December 25, pp. 613–614.

———. 1921h. La crisi commerciale ed industrial nelle sue origini e nelle sue ripercussioni [The origins and repercussions of the commercial and industrial crisis]. *La Perseveranza*, June 19.
———. 1921i. Gli aspetti della crisi bancaria [Features of the banking crisi]. *Il Secolo*, December 31.
———. 1921l. La legge del pane [The bread law]. *Il Resto del Carlino*, March 4.
———. 1921m. Il marco e la crisi germanica [The mark and the crisis in Germany]. *Il Resto del Carlino*, November 17.
———. 1921n. Il bilancio e i cambi [The budget and the exchange rate]. *Il Secolo*, December 20.
———. 1922a. L'avenir de l'Europe. Le point de vue d'un italien [The future of Europe. An Italian's point of view]. *Revue de Genève*, October, pp. 438–448.
———. 1922b. L'uso della forza [The use of force]. *Il Giornale di Roma*, September 24.
———. 1922c. Problemi insolubili [Insoluble problems]. *Il Secolo*, January 18.
———. 1922d. Oggi e un secolo fa [Today and a century ago]. *Il Secolo*, March 25.
———. 1922e. La crisi del risparmio [The savings crisis]. *Il Secolo*, April 15.
———. 1922f. Apparenza e realtà [Appearance and reality]. *Il Secolo*, May 2.
———. 1922g. Cause ed effetti [Causes and effects]. *Il Secolo*, April 20.
———. 1922h. Fatti e commenti [Facts and comments]. *Il Secolo*, July 11.
———. 1922i. La Russia [Russia], *Il Secolo*, June 13.
———. 1922l. Il fascism [Fascism]. *La Ronda*, January, pp. 39–52.
———. 1922m. Il discorso di Treves [Treves' speech]. *Il Secolo*, August 17.
———. 1922n. Crisi bancaria ed industriale ed economia nazionale [The banking and industrial crisis and the national economy]. *La critica politica*, 25 gennaio, pp. 3–11.
———. 1922o. Previsione dei fenomeni sociali [Forecast of social phenomena]. *Rivista d'Italia*, April 15, pp. 370–389.
———. 1922p. Le otto ore di lavoro [The eight working hours]. *Giornale d'Italia*, January 26.
———. 1922q. Verso un risanamento economico [Towards an economic recovery]. *Il Resto del Carlino*, 18.
———. 1922r. I voti dei bancari [Bankers' votes]. *Il Secolo*, June 16.
———. 1922s. Fatti e commenti [Facts and comments]. *Il Secolo*, July 11.
———. 1922t. Economie e spese [Savings and expenditure]. *Il Secolo*, July 29.
———. 1922u. I cambi a Genova [The exchanges at Genoa]. *Il Secolo*, April 23.
———. 1923a. Il trattato ineseguibile [The treaty which cannot be implemented]. *Il Secolo*, February 21.
———. 1923b. La guerra continua [The war goes on]. *Il Secolo*, April 20.
———. 1923c. Francia e Russia [France and Russia]. *Il Secolo*, marzo 4.
———. 1923d. Fascismo e le classi [Fascism and social classes]. *Il Nuovo Paese*, January 3.

———. 1923e. Pareto e il fascism [Pareto and Fascism] interview by Amedeo Ponzone. *La Tribuna*, April 24.
———. 1923f. Paragoni [Comparisons]. *Gerarchia*, January, pp. 669–672.
———. 1923g. El fenómeno del fascism [The phenomenon of Fascism]. *La Nación*, March 25.
———. 1923h. Los partidos politicos [The political parties]. *La Nación*, June 10.
———. 1923i. Pochi punti di un futuro ordinamento costituzionale [A few points concerning a future constitutional system]. *La vita italiana*, September-October, pp. 165–169.
———. 1923l. La crisis del parlamentarismo [The crisis of parliamentarianism]. *La Nación*, September 1.
———. 1923m. Desocupación y depreciación [Unemployment and depreciation]. *La Nación*, June 17.
———. 1923n. I provvedimenti del governo [The government's measures]. *Il Giornale Economico*, January 10, pp. 1–3.
———. 1923o. In margine del bilancio de Stefani [About de Stefani's budget]. *Il Giornale Economico*, June 10, pp. 161–163.
———. 1966. *Mythes et Idéologies [Myths and ideologies]*. Complete Works, tome VI, ed. Giovanni Busino. Geneva: Droz.
———. 1974. *Écrits épars*, OEuvres Complètes, t. XVI, Genève, Droz.
———. 1975. *Epistolario 1890–1923 [Correspondence, 1890–1923]*. Complete Works, tome XIX-2, ed. Giovanni Busino. Geneva: Droz.
———. 1980. *Écrits sociologiques mineurs [Minor sociological writings]*. Complete Works, tome XXII, ed. Giovanni Busino. Geneva: Droz.
———. 1984. *Lettere a Maffeo Pantaleoni 1907–1923 [Letters to Maffeo Pantaleoni 1907–1923]*. Complete Works, tome XXVIII.III, ed. Gabriele De Rosa. Geneva: Droz.
———. 1987. *Écrits politiques. Reazione, Libertà, Fascismo, 1896–1923 [Political writings. Reaction, Liberty, Fascism, 1896–1923]*. Complete Works, tome XVIII, ed. Giovanni Busino. Geneva: Droz.
———. 1989. *Lettres et Correspondances [Letters and correspondence]*. Complete Works, tome XXX, ed. Giovanni Busino. Geneva: Droz.
———. 2001. *Nouvelles Lettres 1870–1923 [New Letters 1870–1923]*. Complete Works, tome XXXI, ed. Fiorenzo Mornati. Geneva: Droz.
———. 2005. *Inédits et Addenda [Unpublished writings and Addenda]*. Complete Works, tome XXXII, ed. Fiorenzo Mornati. Geneva: Droz.
Ritschl, Albrecht. 1999. Les reparations allemandes, 1920–1933: une controverse revue par la théorie des jeux [The German war reparations 1920–1933: a controversy interpreted according to games theory]. *Économie internationale* 2: 129–154.
Spriano, Paolo. 1964. *L'occupazione delle fabbriche. Settembre 1920 [The occupation of the factories, September 1920]*. Turin: Einaudi.

CHAPTER 8

The Final Phase of Paretology During Pareto's Lifetime

To complete Pareto's intellectual biography, we may consider it sufficient to provide an overview of Paretology as represented in publications dating to the final period of his life. In this regard it may be stated, paradoxically, that the two works which have most contributed to his lasting fame aroused less interest at the time of their publication than did the *Cours d'économie politique*.[1] This fact can be explained by the numerous conceptual novelties contained in both the *Manual of Political Economy* (which also displayed a high level of formal complexity[2] by the standards of the time, even if this was mostly limited to the appendix) and the *Treatise on General Sociology* (which furthermore was published in the middle of the world war). In the following sections we will summarise what we consider the most interesting among the responses to the *Manual* (§1), to *Socialist systems* (§2) and to the *Treatise* (§3).

8.1 The *Manual of Political Economy*

To begin with, the most important and interesting episode in Pareto studies within the economic field, after the publication of the *Course in political economy*, was the amicable methodological dispute between Pareto and Croce at the beginning of the century.[3]

Croce, after first praising Pareto for having distanced economics from the historical and empirical spheres, thereby elevating it to the status of a science, stated nevertheless that he did not approve of the mechanistic,

F. Mornati, *Vilfredo Pareto: An Intellectual Biography Volume III*, Palgrave Studies in the History of Economic Thought, https://doi.org/10.1007/978-3-030-57757-5_8

hedonistic, technological and egotistic interpretations given by Pareto, although he acknowledged Pareto's having later repudiated the first two of these.[4] For Croce, rather, "economic acts [are] practical acts, as they are emptied, through abstraction, of all content, moral or immoral".[5]

Pareto first accepted Croce's definition, adding only that he wished to investigate the patterns relating exclusively to the specific practical activities associated with the production and exchange of goods.[6] In performing these investigations, he made use of the methodology of the physical sciences, that is by making reference to increasingly concrete notions linking the observed phenomena. Thus, after having initially adopted the hedonist hypothesis and having found it unsatisfactory due to the non-measurability of pleasure, he turned to "the material fact of choice", in the expectation however that "one day someone else will come along and find something better and simpler".[7] Having said all this, Pareto thought that in Croce's definition of economics it was "not easy to define the term practical", suggesting that for economics, too, one would have to be satisfied with "reasoning with no clear idea of the confines".[8] As an alternative, he proposed the treatment of economic actions as logical[9] in that choices, when purposeful, are structured in a logical manner.[10]

In his reply, Croce pointed out that, paradoxically, it was Pareto himself who was reasoning in a metaphysical manner, with his insistence (a prejudice typical of mathematicians[11]) that "the characteristics of human activity must be of the same nature as physical characteristics", whereas in fact the former are "internal manifestations", while only the latter are "external manifestations", or phenomena.[12] In giving his final response, Pareto claimed to "be the most nominalist of the nominalists", considering that, "objectively speaking", only concrete phenomena exist, whose classification can in fact be deduced logically, if only on the basis of arbitrarily-determined aims.[13] There is no difference between the uniformity of physical phenomena and that of human phenomena.[14] Be this as it may, a final chapter in this unlikely debate was seen in March 1906 when Croce, in a review of the *Manual of Political Economy*, reiterated his acknowledgement of Pareto's role in freeing economics from its political, practical and philosophical trappings and assigning it the task of investigating economic phenomena in themselves, in accordance with the increasingly rigorous methodology provided by mathematics.[15] Nevertheless, having said this, he charged Pareto with seeking to deny others the right to engage in the philosophy of economics which, in Croce's view, was the only method which could "permit the comprehension" of economic phenomena.[16] For, according to the reviewer, this was indeed Pareto's design in presenting his gnoseology and his ethics in the first two chapters of the *Manual*, which, owing to Pareto's lack of relevant expertise, contained "the most speculative and unjustifiable assertions", such as that purporting to characterise all non-empirical propositions as being unscientific, or false.[17] Pareto replied privately to Croce,[18] saying that "the difference [between them] was that [for Pareto] concepts and terminology are secondary in comparison with the things they refer to [, while for Croce concepts and terminology] are the things themselves".[19]

Among the reviews of the *Manual* he examined, Pareto most favoured[20] the one by his student Guido Sensini.[21] Here, Sensini characterised the first chapter of the volume as being "a kind of general introduction to the scientific study of any class of phenomena",[22] before underlining the importance of the second chapter for its exposition of "an analysis of human actions and sentiments which immediately appears indispensable for any field of social studies".[23] He went on to say that Pareto, now free of the didactic requirements of the *Cours*, was able to articulate economic theory no longer as the theory of production and exchange but through the tripartite division into the theory of preferences, the theory of obstacles and the theory of general equilibrium,[24] with the latter characterised as the intersection between the equilibrium of preferences and the equilibrium of obstacles.[25] Among the innovations advanced by Pareto, Sensini highlights the substitution of the "relatively unrigorous notion of capital" with the "much more precise one of transformations of economic assets",[26] together with the demonstration that economic theory "can also dispense with price".[27]

The mathematical aspects of the work were addressed in the review by the mathematical physicist Vito Volterra and, more especially, by Luigi Amoroso and by the American mathematician Edwin Wilson. Volterra, after reiterating the importance of the transition from the notion of ophelimity to that of the curve of indifference,[28] expressed his hope for a fuller treatment of the case involving two commodities[29] and his appreciation of Pareto's use of diagrammatic representations[30] for the manner in which it is able to render certain questions, "complex and challenging at first glance", more accessible. As for Amoroso, retracing the economic theory developed by Pareto in the *Manuel*, he maintained that it is based on the concept of index functions, that is a function whose value increases (decreases) with increases (decreases) in the individual variables, to describe a state of equilibrium when all of its partial derivatives are null.[31]

On the other hand, Wilson[32] thought that the *Manuel d'économie politique*[33] "is not itself primarily mathematical except in spirit".[34] Examining the appendix, he noted that the relative prices yielded by the surface of indifference are not market prices "but the prices which the given individual would himself establish",[35] with the economic agent exploiting this difference to undertake exchanges which will continue until he has maximised his well-being or until his prices are equal to market prices.[36] Pareto's discussion of the order of consumption could be ignored, because it involved time, which, by definition, did not form part of the static analysis to which Pareto had confined his attention.[37]

The most detailed and meticulous analysis came from Wicksell, whose critique included the contention that the definitions of maximum ophelimity and of free competition corresponded, while emphasising that in this type of market, paradoxically, maximum ophelimity can also obtain when "a huge profit for the few" compensates the "lasting pauperism of the greater part of the population",[38] together with an acknowledgement of Pareto's demonstration, although not formulated "with due care", that free competition associated with private property leads to the highest level of production technologically possible, which would in theory arise in a collectivist society[39] and an objection to the failure to allow for the time factor in the theory of production.[40] In the final analysis, Wicksell, while appreciating Pareto's talents, felt that he had not brought about any revolution in economics.[41]

In Italy, too, there was no lack of explicit criticism of the interpretation of economic theory pursued by Pareto. This included in particular Pasquale Iannaccone's claim that "with the same logical rigour" as Pareto's analysis of economic equilibrium as it appears in the *Manual*, various models of economic science (and hence not only of the mechanistic type) could be constructed from the same representations of reality, each quite legitimately making use of its own family of concepts.[42]

8.2 *Systèmes Socialistes*

Systèmes Socialistes had a more animated critical reception than the *Manual*, no doubt due to the interest generated by the recent *Cours d'économie politique*. According to Pasquale Boninsegni,[43] the starting point for Pareto's investigations was that the imperfection of the social institutions led people to seek causes for society's ills, invariably seeking them, however, in simplistic factors, leading to an often fanatical advocacy of social theories, despite their complete lack of scientific corroboration. The last of these simplistic social theories was Marxism, which differed only superficially from its various and cognate religious forebears.

Boninsegni then dwelt specifically on Pareto's thesis that social reformers make the mistake of believing that mankind can be considered separately from his social institutions. Instead, the truth is that social systems are born precisely out of the interaction between human beings and institutions, making it illusory to expect to convert one type of society to another simply through an arbitrary variation in one of the elements of the social system itself (to wit, the institutions). Moreover, the reformers are incapable of giving clear definitions of the institutional modifications they propose. This is the case, in particular, with Bastiat (who never clearly defines what is meant by the economic liberty which he claims as a panacea for all social ills) but also with Walras himself, who never specifies the confines between individual and state activities, nor how these can be reconciled.

In his brilliant review of the first volume, Pareto's friend Francesco Papafava[44] was immediately struck by Pareto's theory that social justice exists only in the sense that a political élite is replaced when it ceases to be of service to society.[45] From this he infers the implication that socialism "concerns itself more with the good of the individual than does individualism",[46] but is not able to achieve this end for the basic reason that it denies the social importance of the function of savings, which can be undertaken exclusively by capitalist individuals.[47] While reviewing the second volume, Papafava, having pondered Pareto's claim that only the form[48] of the struggle for survival is susceptible to change, finally agreed that this struggle is indispensable for progress.[49] An important example of this type of struggle is class struggle, which will bring only a small proportion of the workers to power.[50] Hence, for Papafava, the book is a "treatise on the philosophy of force".[51]

Georges Sorel, for his part, having complimented Pareto on meeting "the need for a complete account of contemporary socialism",[52] expressed his agreement with the idea that "the historical value of Marxism is completely independent of the value of *Das Kapital*" and consisted in its capacity to give "the working class the sensation of power".[53]

The list of the work's critical reviewers began from Croce who, identifying within Pareto's book "two of the supporting pillars of Marx", which were "the idea of class struggle and the lies and slogans which are used, often unwittingly, to disguise conflicting economic interests"[54] and recognising that the work contains "a wealth of doctrine and of perceptive and original ideas",[55] went on to wonder if the "series of observations" offered by Pareto can be "considered as social science or as out-and-out science".[56] Indeed, contrary to Pareto's belief, the factor hindering the creation of a social science is not sentiment, which an academic is able to shield himself from, but the fact that "history and society are the domain not of the intellect but of intuition" since they are characterised, in Aristotelian terms, not by the necessary but by the contingent. Further, intuition alone permits us to arrive at a knowledge of the truth even in regard to social matters, represented by a "potentially accurate vision of the real circumstances and forces which can be harnessed to modify them". Having said this, the reviewer expressed his agreement "with the greater part of the factual observations and the specific concrete arguments" contained in the book, which he calls "excellent".[57]

According to Giovanni Vailati, Pareto's book aimed to investigate the reasons for the success or otherwise of the various socialist systems seen during the course of history.[58] Among the principal explanations, Pareto identified the turnover of élites and, in addressing this, he particularly emphasised the "complete and utter" contrast between the egalitarian formulae which were explicitly declared by an élite during their struggle for power and the implementation of these in reality.[59] Vailati observed in this regard, however, that in many revolutions "some part, however small, of the idealistic aspirations which contributed to their advent"[60] had been achieved. Moreover, the reviewer thought that "the reforming or revolutionary tendencies" of the time were too distinct from those of past societies to allow the formulation of "deductions or speculations" derived from the study of the latter, as Pareto did.[61] On the other hand, according to the reviewer, Pareto gave the best of himself as a "subtle psychological observer and acute analyst of the various forms of fallacy and illusion" to which "inventors of ideal forms of society" fell victim, particularly the tendency to attribute social ills to given institutions and not to human nature,[62] failing to draw up a full evaluation of the positive and negative aspects[63] of existing or desired institutions, ignoring the question of the choice of those to be given power, which was a matter of fundamental importance when, as for any scheme of social reform, competition is eliminated as the main mechanism of selection for future leaders.[64]

Charles Gide, on the other hand, thought that Pareto's aim had been to "draw attention to the falsehoods promoted by the socialists",[65] with the sole exception of Marxism, because, firstly, it encouraged "the abandonment of the habit of explaining facts by reference to ideas" and secondly because of its celebration of class struggle as the instrument of the requisite process of social selection.[66] On the other hand, the reviewer considered Pareto's claim that the various social reforms had not met with the least success in improving "individuals of inferior races"[67] to be exaggerated and, in general, noted a certain disorganisation in the work's presentation, which, however, he considered the "beauty spot" which is characteristic of those possessing an excessive abundance of ideas.[68]

8.3 The Treatise

In what Pareto believed to be the best review of the work,[69] Giovanni Papini[70] said that "the most important characteristic of Pareto's thinking is that it is non-religious. Let us be clear: non-religious does not mean anti-religious" and this afforded him a "freedom of judgement whereby he was able to see the consistencies and the inconsistencies of all faiths and doctrines" while personally seeking to avoid the "pitfalls of sentiment and of language". In this context, Pareto thought that through Pareto's *Treatise* alone, particularly in the final two chapters, "sociology passes from the theological-metaphysical arena to the scientific arena". In Papini's view, the book can thus be criticised effectively only by demonstrating that "Pareto's initial postulates (the differentiation between sentiments and arguments, observation as the sole credible basis for science) contrast excessively with experience" or (based on differing evidence from that invoked by Pareto) that "most human actions escape the classifications" developed by Pareto.

The ranks of the critics, some decidedly virulent, were led by Achille Loria[71] who, finding it incongruous that "a book promoting rational / empirical science" should be dedicated almost exclusively to the consideration of actions which people are prompted to perform by inexplicable impulses,[72] stated that in the *Treatise*, "the author's so-called rational/ empirical science leaves only a very secondary role, as an afterthought, to logic", with priority given to "nude and crude experience".[73] More specifically, if, as Pareto claimed, logic, in formulating derivations, "serves subjective and partisan ends", he himself could also have been guilty of the same transgression.[74] Moreover, and more particularly, by declaring non-logical actions to be inexplicable, Pareto, the anti-metaphysician, paradoxically arrived at the same conclusions as Henri Bergson, the declared enemy of rational/empirical science.[75] This position of Pareto's, which Loria disputes on the basis that he considered that non-logical actions were simply actions which individuals perform in their own interests but which give rise to "completely different and unexpected consequences",[76] was also the reason "for the absolute pre-eminence" which Pareto ascribed to religious factors.[77] Basing himself on the Pareto's own very copious documentation on the failings of leadership, Loria also challenged the significance of the theory of élites.[78]

Also of interest is the spirited review by the socialist scholar Franz Weiss, which begins by accusing Pareto of (1), displaying "[a] presumption and [an] arrogance which extend too far beyond the true, relatively mediocre, scientific value of the book", (2) of being "very far removed from the serenity of spirit and the lofty all-encompassing vision" indispensable in order to "perform productive work in the field of social studies",[79] (3) of being anything but rigorous in his application of the rational/empirical method which, in any case, was far from being a novelty[80] and (4), of ignoring the two authors, Hegel and Vico, who had preceded him in making attempts at producing a general sociology.[81]

After this, the reviewer affirmed that Pareto's "spiritual philosophy" amounted only to the assertion that the majority of human actions are non-logical, while the explanations given (i.e. the derivations) were "insufficient and extraneous".[82] The first of these assertions is correct but, in expounding it, Pareto committed the two capital errors of not distinguishing between the different types of non-logical thought (i.e. faulty reasoning and plain non-reasoning) which lie behind non-logical actions[83] and ignoring the fact that the idea of non-logical actions had already appeared both in Vico and in Hegel.[84] Likewise, the second assertion, traceable to the fundamental debate concerning the means of distinguishing between truth and error, was correct but Pareto's exegesis of it was extremely lacking in clarity compared to Hegel's.[85] On the other hand, Weiss acknowledged that Pareto had been the first to address "the practical tendencies" alternative to the "economic process", although his exposition of residues appeared "chaotic and incomplete". Specifically, he wondered why Pareto had not also included economic interests.[86] Lastly, the reviewer asserted that the criticism Pareto appeared to address towards all social classes was in reality "exquisitely tendentious and reactionary" in that it had the aim of discouraging participation in the "conflict between the bourgeoisie and the proletariat".[87]

Luigi Einaudi, for his part, after commenting on the scathing character of the writing and the harshness of the critical judgements expressed, identified the definitions and the illustrations of residues and derivations as the kernel of the work. Then, pointing out that the nature of social phenomena changes in line with alternations in the relative importance of the two principal types of residues (the instinct for combinations and the persistence of aggregates), Einaudi expressed the view that what Pareto defined as "the oscillations and the undulations" of society was referred to the same phenomena which Vico had termed the "ebb and flow of history".[88]

The sociologist Célestin Bouglé, describing Pareto's basic thesis as being the idea that "many things which are not logically valid are nevertheless socially advantageous", added that "[the work] is difficult to follow [and] the author's ideas remain impenetrable", particularly as he had omitted to correlate them with those of authors such as Durkheim, Lévy Bruhl or Ribot who had covered the same ground not long since. All in all, it was a book by an "original" which itself was anything but original.[89]

The sociologist Maurice Halbwachs, having acknowledged Pareto's "extensive erudition and his painstaking analysis and classification",[90] criticised his failure to discriminate between experimentation and exemplification, which seemed designed to "identify an instinct which, a priori, seemed [to Pareto] an essential and constant feature of human nature",[91] his lending the distinction between logical and non-logical actions an importance which might obtain in the physical sciences but which is certainly lost in social science, where it corresponds to the merely ideological distinction between actions conforming or otherwise to social equilibrium[92] as well as his having adopted an arbitrary approach to the comparison of the elements from which he had construed the residues.[93]

Finally, Croce, who considerately delayed the publication of his review until after Pareto's death, described the *Treatise* as "a case of scientific teratology",[94] which was inevitable in an attempt to study society via a positivist methodology. In particular, Croce challenged Pareto's interpretation of actions not as emerging from the spirit but as objective phenomena, leading to a conception based on ideas which were not new but only vague, such as the distinction between logical and non-logical actions. The only interesting aspect of the book, although itself only superficial and not innovative, "is the characterisation of force as the driver of political events".[95]

Notes

1. For a survey of the critical fortunes of the *Cours*, see (Mornati 2018 chapter XIII).
2. This was emphasised, albeit with courtesy and admiration, in the review by Philipp Wicksteed, "The Economic Journal", 1906, 64, pp. 553–557: 554.
3. In regard to this, one of the best-known episodes in Pareto studies, see recently (Montesano 2003) and (Silvestri 2012, pp. 55–95).
4. See (Croce 1900, p. 16).
5. Ibid., p. 24.
6. See (Pareto 1900, p. 140).
7. Ibid., pp. 143–144.
8. Ibid., p. 161.
9. Ibid.
10. Ibid., pp. 155–156.
11. See (Croce 1901, p. 130).
12. Ibid.
13. See (Pareto 1901, p. 131).
14. Ibid.
15. See (Croce 1906, reprinted in Croce 1977, p. 241)
16. Ibid., p. 242.
17. Ibid., p. 243.
18. Again, in the *Manuel*, Pareto mentions some of Croce's latest objections only to relegate them to the status of manifestations of the metaphysical approach, which was perfectly valid but which he himself did not choose to adopt, see (Pareto 1909, chapter I, §10, note 1).
19. Pareto to Croce, 21st March 1906, see (Pareto 1975a, p. 563).
20. Pareto to Sensini, 20th May 1906, see (Pareto 1975a, p. 565).
21. See (Sensini 1906).
22. Ibid., p. 174.
23. Ibid.
24. Ibid., p. 179.
25. Ibid., p. 192.
26. Ibid., p. 190.
27. Ibid.
28. See (Volterra 1906, p. 299).
29. Ibid., p. 300. As is known, Pareto indulged Volterra's wish, see (Pareto 1906) and Pareto (1909, appendix, §13).
30. See (Volterra 1906, p. 301).
31. See (Amoroso 1909, p. 354).
32. Who inspired Paul Samuelson's economics, see (Carvajalino 2019).

33. Which he reviewed following a wish expressed by Pareto in the letter to Wilson of 16th September 1908, see (Pareto 1975a, p. 368).
34. See (Wilson 1912).
35. Ibid., p. 467.
36. Ibid., p. 469.
37. Ibid., p. 468.
38. See (Wicksell 1913, reprinted in Wicksell 1958 p. 168).
39. Ibid.
40. Ibid.
41. Ibid., p. 175.
42. See (Jannaccone 1910).
43. See (Boninsegni 1903).
44. See (Papafava 1902), a review which Pareto considered excellent, Pareto to Papafava, 30th September 1902, see (Pareto 1975a, p. 413).
45. See (Papafava 1902, p. 86).
46. Ibid.
47. Ibid., pp. 88–90.
48. See (Papafava 1903, p. 460).
49. Ibid., p. 461.
50. Ibid., p. 463.
51. Ibid., p. 467.
52. See (Sorel 1903, p. 220).
53. Ibid., p. 222.
54. See (Croce 1902).
55. Ibid.
56. Ibid.
57. Croce wrote this review, which limited itself to the first volume, on Pareto's prompting, Pareto to Croce, 10th and 12th June 1902, see (Pareto 1975a, p. 461).
58. See (Vailati 1903, 286).
59. Ibid., pp. 286–287.
60. Ibid., p. 287.
61. Ibid., p. 288
62. Ibid.
63. Ibid., p. 289.
64. Ibid., p. 292.
65. See (Gide 1903, p. 168).
66. Ibid., p. 170.
67. Ibid.
68. Ibid., p. 171.
69. Pareto to Sensini, 5th April 1917, see (Pareto 1975b, p. 958). Pareto particularly appreciated the fact that here he was viewed as "an atheist with

regard to all religions, including metaphysics, but an atheist who nevertheless recognises the great value which … religions can have for society", Pareto to Sensini, 23rd April 1917, ibid., p. 961. Pareto also expressed his approval of Sensini's synopsis, see (Sensini 1917) Pareto to Sensini, 23rd April and 3rd September 1917, see (Pareto 1975b, pp. 960, 985); Pareto to Pantaleoni, 31st August 1917, see (Pareto 1984, p. 215).

70. See (Papini 1917).
71. See (Loria 1917).
72. Ibid., p. 377.
73. Ibid., p. 378.
74. Ibid.
75. Ibid., p. 379.
76. Ibid., p. 380.
77. Ibid.
78. Ibid., p. 382.
79. See (Weiss 1917a, p. 138).
80. Ibid.
81. Ibid., p. 140.
82. See (Weiss 1917b, p. 165).
83. Ibid.
84. Ibid., p. 166.
85. See (Weiss 1917c, p. 177).
86. Ibid., p. 179.
87. See (Weiss 1917d, p. 188).
88. See (Einaudi 1917). Pareto thought that "Einaudi's is the worst [review]" precisely because he was "placed with Vico!", expostulating that it was "just like putting water with fire. An enemy of metaphysics with the paramount metaphysical genius like Vico!", Pareto to Bodio, 17th April 1917, in Pareto, *Nouvelles lettres (New letters)*, p. 297; but also Pareto to Pansini, 23rd April 1917, see (Pareto 1989, p. 583); Pareto to Placci, 28th June 1917, see (Pareto 1975b, p. 973).
89. See (Bouglé 1919, p. 128).
90. See (Halbwachs 1918, p. 580).
91. Ibid., p. 581.
92. Ibid., p. 582.
93. Ibid., p. 584.
94. See (Croce 1924, p. 172).
95. Ibid., p. 173.

Bibliography

Amoroso, Luigi. 1909. La teoria dell'equilibrio economico secondo il prof. V. Pareto [The theory of economic equilibrium according to Prof. V. Pareto]. *Giornale degli Economisti e Rivista di Statistica* XX (XXXIX-4): 353–367.

Boninsegni, Pasquale. 1903. I sistemi socialisti [Socialist systems]. *La Vita Internazionale, pp.* 9–12 (41–44): 115–118.

Bouglé, Célestin. 1919. 'Traité de sociologie generale' de Vilfredo Pareto [Vilfredo Pareto's treatise on general sociology]. *Revue historique* 1: 128–129.

Carvajalino, Juan. 2019. Edwin B. Wilson, more than a catalytic influence for Paul Samuelson's foundations of economic analysis. *Journal of the History of Economic Thought* XLI (1): 1–25.

Croce, Benedetto. 1900. Sul principio economico. Lettera al Professor Vilfredo Pareto [On the principle of economics. Letter to professor Vilfredo Pareto]. *Giornale degli Economisti* XI (XXI-1): 15–26.

———. 1901. Sul principio economico. Replica all'articolo del Professor Vilfredo Pareto [On the principle of economics. Reply to professor Vilfredo Pareto's article]. *Giornale degli Economisti* XII (XXII-2): 121–130.

———. 1902. Recensione dei Systèmes Socialistes di Pareto [Review of Pareto's Les Systèmes Socialistes]. *Il Marzocco*, 27, July 6.

———. 1906. Economia filosofica ed economia naturalistica [Philosophical and naturalistic economics]. *La Critica* 2: 129–134.

———. 1924. Recensione al 'Trattato di Sociologia generale' di Vilfredo Pareto [Vilfredo Pareto's Treatise on general sociology]. *La Critica* 3: 172–173.

———. 1977. *Materialismo storico ed economia marxistica [Historical materialism and Marxist economic].* Bari-Rome: Laterza.

Einaudi, Luigi. 1917. Il 'Trattato di sociologia generale' di Vilfredo Pareto [Vilfredo Pareto's 'Treatise on general sociology']. *Corriere della Sera*, February 26.

Gide, Charles. 1903. Les systèmes socialistes [Socialist systems]. *Revue d'économie politique*, pp.168–171.

Halbwachs, Maurice. 1918. Le 'Traité de Sociologie générale' de M. V.Pareto [Prof. Vilfredo Pareto's 'Treatise on general sociology']. *Revue d'économie politique*, pp. 578–585.

Jannaccone, Pasquale. 1910. Alle frontiere della scienza economica [On the frontiers of the science of economics]. *La Riforma sociale*, pp. 18–42.

Loria, Achille. 1917. Nuove correnti in sociologia [New tendencies in sociology]. *Nuova Antologia*, pp. 376–382.

Montesano, Aldo. 2003. Croce e la scienza economica [Croce and the economics]. *Economia politica* 2: 201–224.

Mornati, Fiorenzo. 2018. *An intellectual biography of Vilfredo Pareto, I, from science to liberty (1848–1891).* London: Palgrave Macmillan.

Papafava, Francesco. 1902. Il nuovo libro di Vilfredo Pareto [Vilfredo Pareto's new book]. *Giornale degli Economisti* XIII (XXV-1): 84–92.

———. 1903. Il secondo volume dei "Systèmes socialistes" del Pareto [The second volume of Pareto's "Systèmes socialistes"]. *Giornale degli Economisti* XIV (XXVII-5): 460–469.

Papini, Giovanni. 1917. Vilfredo Pareto. *Il Resto del Carlino*, January 21.

Pareto, Vilfredo. 1900. Sul fenomeno economico. Lettera a Benedetto Croce [On economic phenomena. Letter to Benedetto Croce]. *Giornale degli Economisti* XI (XXI-2): 139–162.

———. 1901. Sul principio economico [On economic principle]. *Giornale degli Economisti* XII (XXII-2): 131–138.

———. 1906. L'ofelimità dei cicli non chiusi [The ophelimity of non-closed cycles]. *Giornale degli Economisti* XVII (XXXIII-1): 15–30.

———. 1909. *Manuel d'économie politique [Manual of political economy]*. Paris: Giard et Brière.

———. 1975a. *Epistolario 1890–1923 [Correspondence, 1890–1923]*. Complete works, tome XIX-1, ed. Giovanni Busino. Geneva: Droz.

———. 1975b. *Epistolario 1890–1923 [Correspondence, 1890–1923]*. Complete works, tome XIX-2, ed. Giovanni Busino. Geneva: Droz.

———. 1984. *Lettere a Maffeo Pantaleoni 1907–1923 [Letters to Maffeo Pantaleoni 1907–1923]*. Complete works, tome XXVIII.III, ed. Gabriele De Rosa. Geneva: Droz.

———. 1989. *Lettres et Correspondances [Letters and correspondence]*. Complete works, tome XXX, ed. Giovanni Busino. Geneva: Droz.

Sensini, Guido. 1906. Recenti progressi delle scienze sociali per opera del Prof. Vilfredo Pareto [Recent progress in the social sciences accomplished by Prof. Vilfredo Pareto]. *La Riforma Sociale* 1906: 173–198.

———. 1917. La sociologia generale di Vilfredo Pareto [Vilfredo Pareto's general sociology]. *Rivista Italiana di sociologia*, March–June, pp. 198–253.

Silvestri, Paolo. 2012. *Economia, diritto e politica nella filosofia di Croce. Tra finzioni, istituzioni e libertà [Economics, law and politics in the philosophy of Croce: Appearances, institutions and liberty]*. Turin: Giappichelli.

Sorel, Georges. 1903. Compte-rendu de Les Systèmes Socialistes [Report on Les Systèmes Socialistes]. *Revue philosophique de la France et de l'étranger*, pp. 220–223.

Vailati, Giovanni. 1903. Recensione dei Sistemi socialisti [Review of Socialist Systems]. *La Riforma sociale* 4: 285–296.

Volterra, Vito. 1906. L'economia matematica ed il nuovo Manuale del Prof. Pareto [Mathematical economics and Prof. Pareto's new manual]. *Giornale degli Economisti* XVII (XXXII-4): 296–301.

Weiss, Franz. 1917a. Il nuovo verbo delle scienze del prof. Vilfredo Pareto [Prof. Vilfredo Pareto's new version of science]. *La Critica sociale* 10: 138–140.

———. 1917b. Il nuovo verbo delle scienze del prof. Vilfredo Pareto [Prof. Vilfredo Pareto's new version of science]. *La Critica sociale* 12: 165–167.

———. 1917c. Il nuovo verbo delle scienze del prof. Vilfredo Pareto [Prof. Vilfredo Pareto's new version of science]. *La Critica sociale* 13: 177–179.

———. 1917d. Il nuovo verbo delle scienze del prof. Vilfredo Pareto [Prof. Vilfredo Pareto's new version of science]. *La Critica sociale* 14: 187–190.

Wicksell, Knut. 1913. Vilfredo Pareto's Manuel d'économie politique. *Zeitschrift für Volkswirtschaft, Sozialpolitik und Verwaltung* 1: 132–151.

———. 1958. *Selected papers in economic theory*. London: Allen and Unwin.

Wilson, Edwin Bidwell. 1912. Recension du Manuel d'économie politique [Review of the Manuel d'économie politique]. *Bulletin of the American Mathematical Society* 18 (7): 462–474.

Epilogue

In this volume we have adopted an inductive methodology, drawing on the whole corpus of Pareto's writings in order to recreate the final quarter century of his life, when he was able to gain his considerable, and continuing, reputation in the field of economics as well as his more controversial, but not negligible, status in that of sociology.

Pareto's conception of theoretical, or "pure", economics, appears to have undergone a definitive, largely self-sustained, development, allowing him to achieve his long-pursued aim of freeing general equilibrium from metaphysical impurities (in particular, the unmeasurable notion of ophelimity).

In any case, the practical aspects of economics continued to occupy the attention of Pareto the applied economist, even if by this time he was convinced that these could be sufficiently understood from a sociological viewpoint. Thus, in the period in question, further investigations were conducted into ideas which he had already explored in greater or lesser depth, such as the theory of international trade, income distribution, economic crises, the demographic phenomenon, progressive taxation and social welfare. However, there were also other, new, topics such as the maximum ophelimity of a community in sociology, an outline of the sociological conception of savings, the statistical study of the relations between economic and social phenomena, the management of public debt, the possible management of the Italian railways by a union and, finally, strikes.

F. Mornati, *Vilfredo Pareto: An Intellectual Biography Volume III*, Palgrave Studies in the History of Economic Thought, https://doi.org/10.1007/978-3-030-57757-5

Drawing largely but not exclusively on his *Systèmes Socialistes*, we then observed, through Pareto's eyes, the apparently unstoppable advance of socialism from the beginning of the century until the eve of the First World War, at the same time highlighting the extremely prescient insight whereby he saw in a great war the only possible obstacle to the final triumph of socialism.

It was, if not expressly, then also, in order to understand these political developments, inaccessible to traditional sociology with its baggage of political ideology, that Pareto produced his rather contorted *Treatise on General Sociology*, conceiving of it explicitly as a "necessary complement to studies of political economy". We have explored the principal innovations of this highly complex work, including the theory of action, social heterogeneity, social equilibrium and some aspects of a sociology of politics, followed, for the record, by an appendix covering the various convoluted publishing projects which culminated in the *Treatise*.

The fortuitous arrival of the world war furnished an opportunity for Pareto to perform a preliminary appraisal of the methods of sociological analysis he had developed in the *Treatise* through their application in the wartime context, in regard to both the international and the Italian scenes.

A similar appraisal was made possible by post-war circumstances, which Pareto observed from the standpoint of his view that "the demagogic plutocracy achieved a total victory" in the recent war and that "the battle between this force and outright demagogy" had now begun, with unforeseeable consequences. This further confrontation, which in Pareto's view would be decided by "force", inspired him to produce a number of further studies which we have reviewed, enumerating them within the categories of a non-economic conceptualisation of social conflict, reflections on the difficulties affecting the League of Nations, the early phases of Bolshevism in Russia and the failed Bolshevik sedition in Italy in the summer of 1920, the beginnings of Fascism, together with the related analysis of the principal economic problems of the post-war period. It is of interest to record that Pareto, struck by the immediately animated character of the controversy surrounding Fascism, thought it wise to provide the explicit assurance that he would "examine Fascism with the absolute impartiality adopted for the analysis of many other political, economic and social phenomena".

We then concluded the volume by reviewing the last period of Paretology during his lifetime, focusing on the reviews of the *Manual*, the *Systèmes socialistes* and the *Treatise*.

We hope, at the culmination of this long investigation based on an inductive methodology, to have substantiated our non-linear interpretation of Pareto's intellectual biography, as represented in the progression from an initial a priori devotion to science, through a long period of commitment to liberal values followed by a renewed and more committed dedication to science as providing the perfect path to existential fulfilment.

Index[1]

[1] Note: Page numbers followed by 'n' refer to notes.

F. Mornati, *Vilfredo Pareto: An Intellectual Biography Volume III*, Palgrave Studies in the History of Economic Thought,
https://doi.org/10.1007/978-3-030-57757-5

Printed in the United States
By Bookmasters